FERN GLEN FARM.

BY

HELEN PINKERTON REDDEN.

With five Illustrations by the Author.

" We cannot buy with gold the old associations."—LONGFELLOW.

London:

HODDER AND STOUGHTON,

27, PATERNOSTER ROW.

MDCCCLXXXIV.

Fern Glen Farm.

TO

MABEL PESTELL

𝔗𝔥𝔢 𝔖𝔱𝔬𝔯𝔶 𝔬𝔣 𝔉𝔢𝔯𝔫 𝔊𝔩𝔢𝔫 𝔉𝔞𝔯𝔪

IS

AFFECTIONATELY INSCRIBED.

That this simple narrative of country life may be a source of pleasure to her, as well as to other young people, and that it may be the means of awakening in their minds a loving interest in the animals around them, is the earnest hope of their sincere friend, the writer.

CONTENTS.

viii   *Contents.*

CHAPTER I.

DESCRIPTION OF FERN GLEN.

CHAPTER I.

DESCRIPTION OF FERN GLEN.

SOME few miles away from any town was a country lane, hedged in on either side by high grassy banks interspersed with silver birch and ash trees. Here and there were low willows, whose gnarled roots spread and jutted out from the bank, forming rustic seats covered with moss and ivy.

In this lane there was every indication of the coming spring. It had been a long and dreary winter, and the lane that was so picturesque and shady in summer had been almost buried in the snow which had drifted there in heaps, and when the thaw set in, it was so furrowed into deep ruts by heavy wains and labouring horses, as to be rendered almost impassable. But now there was a change.

It had advanced so gradually as to be scarcely perceptible, but without doubt fairy-like spring once more had condescended to visit the dark barren earth.

Everywhere could be traced the print of her light footstep. The grass grew once more young and fresh; violets raised their small fragrant heads everywhere; primroses, cool and pale, peeped from the mossy banks beneath the hedgerows, while over all the sun threw his warm, bright radiance.

Birds began to carol forth their sweetest lays after

their long winter silence. The hum of the insect, the bleating of the young lambs, the murmuring of the narrow brooklet in the hollow,—all seemed to unite and to re-echo in one unceasing song of praise to God for the return of spring.

About half a mile to the left was the village, consisting of a few small cottages, besides two or three houses of larger pretensions. The afternoon sun was reflected in dazzling rays in each of the tiny latticed panes of the windows, and shone upon the curly flaxen head of many a small child as it stood in the bright trim garden before the cottage door.

In the midst of the village was the old grey church, with its square tower covered with ivy, which boasted of three bells, whose jangling chimes might be heard wafted on the breeze on a quiet Sunday morning, now softer, now louder, and at length dying away in faint reverberations. Later in the season roses would bloom thickly around the deep Gothic windows.

In the small church-yard amongst the long grass were a few grey moss-covered tombstones, some in the last stage of decay, with their inscriptions and cherubs' heads almost obliterated. Inside the church were high square pews, and hanging in one corner, crumbling to dust, were a suit of armour and a banner or two.

Faded hieroglyphics and strange quaint frescoes decked the walls; one of St. George and the Dragon stood out more prominently than the rest. In summer time the voices of the few worshippers assembled there would often be almost drowned by the united warblings of thrushes and blackbirds, as they sat and sang in the old trees near at hand.

Leaving the village some distance behind, the wind-

ings of the shady lane before mentioned led to a large
old-fashioned farm-house, known as "Fern Glen Farm,"
gable-roofed, lattice-paned. The right and left wings
of the building projected on to a verandah, which
extended the whole length, and was divided in the
centre by the wide hall door.

In summer-time ivy and roses, westeria and honey-
suckle, clothed the front of the house in one glowing
array of bloom and fragrance. The hedge surrounding
the well-kept garden was of sweet-brier ; on the smooth-
shaven lawn were rustic roots filled with flowers, and
in the midst was a fountain encircled by gravel and
pink-lipped shells.

Under some of the trees stood a rustic seat or two,
and down the slope to the river a boat lay moored to
the bank, its motionless reflection mirrored in the clear
water. The fish could plainly be seen wending their
way in and out of the rushes and reeds, and very soon
forget-me-nots, pretty wee pink flowers, and numberless
small offspring of the woods and valleys, with their
tender leaves of delicate green, some serrated, others
variegated with scarlet tips, would be seen dotting the
sedgy banks, as the summer advanced.

The owner of this country residence, with its garden,
river, and surrounding fields, was one Farmer Rothesay,
who lived there with his wife, his two sons and three
daughters, the youngest of whom was the pet of the
household, a little lass of six summers. Her fairy tripping footsteps could be heard at intervals as she danced
singing about the house, up and down the wide oak
staircase, or playing in the hall.

On either side of the door were recesses, in each of
which was a long narrow window, divided from the

hall by crimson curtains. In these recesses the little girl had built quite a grotto of seaweeds and a variety of pretty delicate shells, surmounted by a small figure of a mermaid—the marine productions she had collected herself when at the seaside for a summer holiday.

Sometimes she went into the dairy and watched the milk skimmed and the butter made, or she would, in her way, assist her mother in her various domestic duties.

From early morning until bed-time, little Meeta was always busy and gay, never idle, ever intent on some childish work or other, and her fair hair and merry face might be seen flitting in and out until the old eight-day clock in the kitchen struck four in the afternoon. Then she would run for her white sun-hat and, tying it over her golden curls, would bound away to meet her father and brothers returning to their five o'clock tea.

Through the white park gate, over the rustic wooden bridge—where the rapid brook gurgled in its bed of mossy stones—away into the fields, where the cattle grazed and the lark was singing high overhead—plucking pink-tipped daisies and golden buttercups as she passed, the long grass rustling at each tripping footstep. Soon she saw the three figures, their shadows stretching long and far before them, and in another moment she was in father's arms, listening to his hearty tones of loving welcome.

Now the old house is in sight, and Mrs. Rothesay stands at the door, shading her eyes with her hand, as she gazes expectant across the fields. Presently they are all seated round the table in the large dining-room, with its polished boards and deep French windows filled with plants, the doors of which open on to the lawn before mentioned.

The snowy cloth is spread, and the table laden with all that makes country fare so delightful in the imagination of a town-bred child. Rich thick cream fills the old-fashioned silver cream-jug; there are the new-laid eggs with their amber shells; the fresh butter modelled by Eva into a tasteful basket of fruit and flowers, and surrounded by parsley. There was virgin honey brimming over its white wax cells, and at the head of the table was a ham of Farmer Rothesay's own feeding and curing.

Now as they sit at tea in the deepening twilight, I must give you some description of their personal appearance.

Mr. Rothesay is of middle height, his complexion is fresh and ruddy; his hair, which is turning slightly grey, curls closely about his temples; he has pleasant, merry blue eyes, and is always ready with a joke or jest. Mrs. Rothesay is somewhat *petite;* her hair is dark and her eyes deep grey; her features are regular and refined; she is always mild and cheerful, and is possessed of a very sweet disposition. She takes such a deep interest in all that concerns her children and their pursuits, that to many she appears more like an elder sister than a mother. Though so young at heart, she yet manages all the affairs of her household with a thrift and thoughtfulness to which few can lay claim.

Hubert, the eldest son, is a fine athletic young man of twenty, thoroughly up in all the mysteries of farming and rearing live stock, and his father's right hand on the farm. He can ride any horse, no matter how vicious— whether he has kicking or shying propensities it is all one to Hu, and after a few times no horse that has once felt his firm hand, his gentle yet strict training, but knows at once that he has found his master, and becomes, outwardly at least, a docile and obedient animal.

Hubert is a first-rate shot, too, and can knock down anything within distance, from a whirring partridge to a dodgy rabbit as it scampers helter-skelter to its home in the wood. He is the best skater, swimmer, and cricketer of all his youthful associates.

Roger, his younger brother, is not so much inclined to field sports. He is slender and handsome, with fair hair ; his eyebrows are dark, and his eyes deep grey, like his mother's. He has a mechanical genius, and having once seen any machine which strikes his fancy, he will make a very correct model of it, which delights his mother, and makes him a wonder among his young companions.

It is some miles into the town where Roger goes to school, and with his satchel of books across his shoulder he has to be off very early in the morning. But on his last birthday his kind father and mother gave him a bicycle, to his great delight, and he at once commenced to make himself master of it, which he soon accomplished, and so the distance between his home and the town seemed short in comparison to what it had been when he was obliged to trudge all the way on foot.

Roger was also a studious boy, and he would sit poring over books of adventure until they produced in him such a liking for a roving life as to lead him to hint more than once that he would make the sea his profession. But the mere supposition brought such a cloud over his mother's fond face that he never had the courage to refer to the subject, but was fain to keep all his longings and aspirations to himself.

Of Eva and Alice, the two girls, we shall have more to say hereafter.

CHAPTER II.

MEETA AND HER PETS.

CHAPTER II.

MEETA AND HER PETS.

PERHAPS you may fancy that when her father and brother were away in the fields, and Mrs. Rothesay and her sisters engaged in household affairs, little Meeta would be idle, and perhaps get into mischief, having no one to look after her, as might be the case with some children in her circumstances. But with her it was not so ; rising early in the morning, she would be up and dressed when many another little girl would be in bed.

For Meeta was a busy child, and by seven o'clock she would be running down the yard with a basket on her arm, into the hen-house. Standing on tiptoe, she puts her little hand first into one nest and then into another, and delightedly places the fresh eggs in her basket. Then she would run back again into the large lofty room where the grain was kept. The poultry did not need to be called, the sight of Meeta and her basket was enough. Down flew the pigeons from the dove-cote, the ungainly geese and ducks came waddling up as fast as their wide feet would let them, the fowls chased and dodged each other about in all directions, making a great cackling all the time, and I am sorry to say that some fought and pecked each other in a very sad way indeed, snatching food from those that were weaker and less able to take care of them-

selves.   There were some pretty little chicks, too, with
long fluffy feathers, speckled all over with black spots.

Amidst all this fluttering and chatter, little Floss the
pet dog would caper about, mingling his short sharp bark
with the rest, no doubt thinking it all very fine indeed.   Meeta would laugh merrily as she looked on at
the antics of her unruly family.

Nor did she forget to take some food to one or two
poor sick fowls, which were ensconced in a basket of
hay in the back kitchen.

Sometimes they would get into a skirmish, and be
sadly hurt, but Mrs. Rothesay was always equal to the
emergency.   She had had to set many a broken leg of
poor unlucky fowls that got into the wars; but she did
it so skilfully, and bound it up so tenderly, that Mrs.
Hen was soon trotting about again as contentedly as
if nothing had happened.

By the time Meeta had finished feeding the poultry
breakfast was ready, and Nettie, the old servant, bustled
off to ring the bell that summoned her master from the
field.   Soon all were seated round the table in the
pleasant dining-room.   Sometimes a relative of Mr.
Rothesay would join them when on his way to the town
on market days, and on this particular morning there
might be heard a quick step echoing along the hall, then
a tap at the dining-room door, while a pleasant voice
asked admittance.   On Hubert springing up to open it,
there entered a tall, somewhat elderly man, with grey
hair and a frank countenance, who was the owner of
some mills a few miles distant from Fern Glen.   Mr.
Rothesay returned his friendly greeting, begging him
to be seated.

"Ha !" said he, with a hearty laugh, which was quite

catching, "there is nothing like an early morning drive for whetting one's appetite, to say nothing of fragrant coffee and fried ham."

Mr. Hargraves was a great man among the young people; he loved to have them about him, and take them over his mills, which he did with evident pride and pleasure. It was the greatest treat to the little Meeta, who would stand for hours if she had been allowed to do so, and watch the great water-wheel, or the white-powdered grain falling over the slide to the trough below. As for Roger, the engine-room was his attraction, and Mr. Hargraves declared that one of these days the boy would be bringing out a grand patent for something or other and make himself a name.

At length, with the exception of the two gentlemen, who were busy discussing crops, grain, etc., all were off about their several avocations. There was always plenty to be done in the morning : there was Roger's luncheon to be got before he started for school, and he was very impatient if it was not there to the minute ; kind Nettie never failed to put in some little delicacy in the shape of fruit or pastry. In the meantime little Meeta would run away to visit the pet lambs— through the wide gravelled yard at the back of the house, past the kennels, where big black old Ben, the house-dog, lay blinking sleepily in the sunshine as it flickered in wavering lines up the wall, while Hu's large retriever, Juno, looked eagerly after her as she passed. Soon she reached the entrance to the field, where, enclosed by hurdles, there were three pet lambs. One had a blue ribbon round its neck and was snow-white, while the other two had black knees and boasted red

and green ribbons, that each might be fed in turn. Soon
Alice would be seen, treading cautiously with a pail of
milk, and one by one she lifted them over the hurdles,
for if they were fed all together they would most likely
have upset it in their eager hurry. Very often, when
Alice had finished feeding them, she would leave them
with Meeta, as they were allowed to be out of the pen
when anyone was by. It would have made a very
sweet picture to have seen the little girl as she knelt,
putting her arms about the neck of first one and then
another of her little favourites.

Now we must introduce you to her rabbits. They
were in the shed at the other side of the barn, where
the ricks towered up so warm and dry, forming a
capital shelter from the cold wind. There was the old
rabbit, looking out expectant from the hutch, and there
were fifteen young ones ! such delicate, sweet little
things, as soft as eider-down, some black and white and
brown, and some with black and brown spots, the exact
image of their mother. She stood on her hind legs and
pushed her nose outside the wire, that she might get
the cool fresh leaves as fast as possible, that her
mistress never failed to bring her. But with strangers
she would become quite savage, and look at them
suspiciously out of her large eyes, for fear they should
injure her little ones. Then there was the pony, who,
when he was a wee colt, and before he had any shoes,
came into the back kitchen and trotted right into the
dining-room and round the table before any one could
drive him out.

But Meeta's great pet was Neddy the donkey. She
could saddle and bridle him herself; he would follow
her about the garden and fields everywhere; she

was the only one who could catch him when loose in the close. He would never let any farm lad come near him, but at the sound of Meeta's high childish voice he would set off at once at a round trot in obedience to her call. Then there was the old duck,

Meeta and the pet lambs.—*P.* 14.

Mrs. Sullivan, as she was called; she was a very sensible old duck indeed, for she could draw a little wooden cart that Roger had made for his sister. He had also procured some red braid for harness, which passed round the duck's neck and over her back to fasten her to the cart.

At half-past ten Meeta had to run to her lessons, after her first duty of feeding her live stock. She was a very intelligent child of her age; whatever she did was done in earnest, whether work or play. She could read and write well, and could ply her needle with no little skill for so young a child. But not only this, she could boast of a tiny lace-pillow! The old nurse and servant Nettie, who had lived with Mrs. Rothesay for many years, had taught her the art of making lace in the long winter evenings, when the snow lay thick on the ground and the wind moaned and whistled in sobbing gusts round the house. So by the time spring came round Meeta could make a simple lace edging quite nicely, and it was very pretty to see her as she sat in the French window, her little fingers flying so rapidly over the smartly-beaded bobbins. She could sing, too, when her kind sister played the accompaniment, and her plaintive child's voice never sounded sweeter than when singing a little hymn that Alice had taught her.

As spring advanced, nothing gave Meeta more pleasure than to go bird's-nesting. When I say bird's-nesting, I do not mean that she dragged the pretty nests away from the hedges where she found them, and took the eggs, unmindful of what the birds would feel when they flew back and found their beautiful little nests all gone, upon which they had spent so much labour and patience. Oh no! Meeta would have been shocked if any one had hinted such a thing. She watched so carefully the inhabitants of the woods and fields, that she knew much of their ways and habits. By a rustic stile in the field there was a certain old tree that stretched over the river; it was quite hollow in one part, and Meeta would clamber up, holding by the strong stems of ivy, and look into the mossy hollow, over which the leaves grew so thickly that no one would have noticed the place except they had a pair of very sharp eyes, and had watched a certain little bird that had flown backwards and forwards a great number of times during the day. Meeta would gaze with delight when she had found the nest and had inspected the blue speckled eggs with their tender fragile shells. She would come again and again, and watch the unfledged nestlings until their feathers were grown, and they ventured with their trembling quivering wings on their first flight from bough to bough, putting their small heads first on one side and then on the other, and looking as it seemed with wonder on the great world that opened before them.

CHAPTER III.

EVA. AND ALICE.

CHAPTER III.

EVA AND ALICE.

AND now I must tell you something of Eva and Alice Rothesay. Eva was her mother's right hand, and busied herself with a thousand things that one finds to do on a country farm. Eva had a great love of birds, and had reared five young bullfinches from the nest. One of the farm lads had found the nest, and Eva had coaxed him with a few pence to give them to her. She was indefatigable in her care of them, rising in the night to give them food, and they repaid her by growing into the handsomest little fellows that you could wish to see.

Their plumage of red and grey, and black and white, was admired by every one who saw them. Sometimes they made such a noise that Mr. Rothesay declared they were almost too much for him. Bullfinches have not much music in their note, though they can be taught several pretty tunes, and their call-note is very soft and sweet. From her childhood Eva Rothesay had evinced a great talent for modelling in clay, so much so that when she had left school she went for some months to London to make herself more proficient. She received some valuable hints while there, it is true, but she learned, too, something of the almost insurmountable difficulties which lay in her path, and her master told

her that if she wished to turn her talent to account she
had better lay aside modelling and take to painting, as
it was more remunerative.   When she returned home
she brought with her several works of considerable
merit.   Very soon she had taken possession of one of
the disused rooms in the house, and made it into a
studio, which was regarded by the rest as her special
property, where she could study without being disturbed.
It was a long narrow room, in which were stowed many
things, generally known as lumber.   There was an old-
fashioned carved oak chest, old bridles and saddles
ranged on the walls, and amidst all were antique models
of hands and feet, a.Niobe, a Hercules, an anatomical
statuette, and some Grecian busts on carved brackets.
In one corner of the room stood her easel, her paint-box,
and her portfolio.   The sole window was framed in
with ivy and creepers, and the light was none of the
best.   Nevertheless Eva worked on, in spite of all draw-
backs.   Like many young amateur artists, she had not
the advantages her more fortunate sisters possessed
who lived in cities, where valuable collections of paintings are open to public inspection ; and she found it
most trying, having none to whom she could go for
instruction or advice.   She had quite as much talent,
and more originality perhaps, than others who were able
to go abroad and cultivate their tastes by the study of
all that is most refined and exquisite in sculpture, as
well as the *chef d'œuvres* of the old masters.   The classic
countries of Italy and Greece were but the shadowy
ideals of a dream that Eva never expected to see except
in books or paintings.   In her isolated position she could
only work on, trusting that light would come some day,
and disperse all the misty shadows of her inexperience,

and help her at last to give to others some faint idea
at least of the numberless fancies floating through her
brain. Above all, Eva was only a girl, which was a
stumblingblock not to be got over. Even if she had
met with an artist friend who would have taken an
interest in her, he would no doubt have considered this
fact as an unfortunate state of things. But there was
not any artist friend, or likely to be, and she had much
to discourage her, as her friends never lost the opportunity of remarking that it was waste of time, that there
was nothing to be gained by it, and that if Eva would
take their advice she would confine herself to house-
keeping and stitching, so that some day she would be
fitted to have a house of her own. All the fashionable
young ladies of her acquaintance considered an "esta-
blishment" the Alpha and Omega of existence, and
thought no scheming and planning too arduous, so that
they might at last be favoured with success. Only Mrs.
Rothesay understood and sympathised with her daughter.
There was one thing that Eva did not like, and that was
needlework. Generally young ladies in story-books are
made to do all kinds of almost impossible feats in that
most necessary art; but as to cooking, pastry and cake-
making, and all the mysteries of the kitchen, few could
excel her.

Alice Rothesay was a sweet-looking girl of sixteen,
with fair hair and large soft blue eyes. She had finished
her last year at school, but though she had done with
the monotonous routine and strict discipline of school-
girl life, she by no means considered her education was
completed, nor did she, like some girls, fling her books
aside as if they were part and parcel of a disagreeable
past that had been got over, and no more need be

thought about it. Alice loved study for its own sake, and kept up her French, music, and singing most diligently. As soon as she came home she begged her mother that the education of little Meeta might fall to her share, and the child in her turn loved her young instructress dearly. Alice was one of those calm quiet natures on whom the trifles and worries of everyday life have little or no effect. Living from her childhood in a country village, having but few companions, and seeing so little of the fashionable world, made her shy and retiring in her manners, and none knew how much was passing within that apparently serene and mild exterior. The pastime she loved best was to take her portfolio and box of water-colours, and wander away to some picturesque spot where she could study as long as she wished. The perfect calm suited her, and it was natural, too, that she should love to hold communion with the sweet spirit of nature, that was in such perfect unison with her own. She could paint very prettily, too, on velvet and satin, and when a birthday came round she had always ready some delicately-painted group of wild flowers, or fancy-work specimens of her dexterous pencil or needle. Her sweet voice was ever a charm and solace to her father when he came home tired of an evening, and an incentive to Roger, who possessed his sister's love of music. He had taught himself many tunes by ear, and he could also accompany her when at the piano on the violin or flute. Thus possessing so many resources in themselves, their solitary country life was anything but dull to them, as it would have been to some people in their position.

CHAPTER IV.

THE GRATEFUL GOOSE.

CHAPTER IV.

THE GRATEFUL GOOSE.

EARLY one morning Mr. Rothesay set out to the pasture to visit his cattle. The shining dew shook off in spray from every blade of grass as he brushed past; the delicate waxen blossom of peach and nectarine bloomed on the wall surrounding the garden, and at every waft of the fresh breeze some of the pink and pearl-like leaves fell fluttering down to the earth. There by the gate were the cattle, and exceedingly fat and well-favoured they looked. Some were grazing, some lying down, others were standing knee deep in the river, their image reflected in the clear stream. The foliage of the trees, as they grew thickly by the water's edge, threw over it dark and ominous shadows, and in its depths lay the azure sky with its flecks of cloud. Every small leaf of the trees above was mirrored there, shading the water with its delicate green, which deepened as it neared the margin, except where the trees widened, and the shades were interspersed with quivering glances of light, which flashed along the surface, making every separate ripple sparkle with a silver sheen.

The little moor-hens peeped every now and then from their hidings amongst the weeds that clothed the banks, uttering their peculiar note; the coo of the pigeons could be heard in the wood, and the rookery was alive with raven plumes which fluttered from tree to tree.

The farmer contemplated his cattle with satisfaction beaming in his eye; the gentle Alderney raised her beautiful head and scanned her master for a moment with her large soft eyes, then went on ruminating as steadily as ever.

As he went first to one, and then to another, each turned and gazed lazily at him. Soon he was wending his way homewards, and as he neared the yard, it seemed to him, by the sounds that fell on his ear, that something unusual had taken place among the feathered tribe. Farmer Rothesay quickened his pace, and was just in time to witness a regular fight between a flock of geese in the yard and the rest of the poultry, aided by the dogs, which were joining lustily. Mr. Rothesay saw that the whole flock had set upon one poor goose, and so many were the knocks and pecks he got, that there is no doubt it would have gone very hardly with him but for the timely interference of the good farmer, who soon put the whole flock to the right-about, and stooping down smoothed the ruffled feathers of the poor bird, which seemed half dead with fright. It soon revived, however, and it was strange to see how this bird, about whose species so much has been said in ridicule, seemed as if it would redeem its race from the stigma that rested on it. Though it was a goose, it was neither silly nor ungrateful, for it never forgot the good services rendered it by its master. Mr. Rothesay never entered the yard afterwards but the bird flew towards him, bending its long neck, rubbing itself against him, and showing in every possible way its affection, and was not satisfied until it had received the accustomed notice.

If he were at the other end of the field, the devoted bird would see him, and, flapping its great wings, would flounder over the hedge which divided the yard from the field, and with its neck outstretched run to meet him, following closely at his heels with all the fidelity of a dog. It would fly at any unlucky person who ventured too near its master, and when he stopped to say a word to any of his men, the goose would quietly squat down on the road and wait until he went on again, when it would rise and follow as before. But it fared ill with the poor goose, for however wise or grateful it might be, its name was not to be got over, and Mr. Rothesay did not altogether like its following him about everywhere ; so the end of it was he sold the bird and saw it no more.

One fine afternoon, when Eva had finished her household duties, she ran upstairs to put on her riding-habit, as Mrs. Rothesay wished her to get a few things from the town. Eva was quite a market woman, and a clever horse-woman besides ; her pony Beauty was not so very easy to ride either ; he had a great deal of corn, which made him spirited ; he was very fat too, and did not always wish to carry his mistress, and he showed his ill-temper by shying at everything he met or saw by the roadside. He was a very handsome little fellow, of a bright bay; his mane was so thick and long that it hung over his eyes, and his tail nearly reached the ground. His knees and hocks were black, shaded on the inner side with light tawny hair. Eva saddled and bridled him herself, as was her wont, and having fastened her marketing-bag to the pommel, she was soon cantering away across the meadow down to the park gate. It was a lovely day, the larks were singing

all around her, and at intervals she could hear the
soft mellow note of the cuckoo in the distance.

The fresh exhilarating breeze roused her, and gave
her for the moment the sense of a wild exultation of
perfect freedom. Giving her pony his head, she was
soon going full gallop down the road. Her eyes
sparkled with animation, and a rich bloom mantled on
her cheek, and by the low thatched cottage doors the
women looked up from their lace pillows and gazed
after the fair equestrian as she passed. The town in
sight, Eva gathered up the reins and proceeded more
quietly. After having accomplished her errands, she
turned her pony's head homewards, and had not gone
far before she espied some one coming in the distance; she could see the habit flying in the breeze
distinctly.

As the figure approached, she knew it to be that of
her friend Mildred Haverleigh, of whose family we shall
have more to say presently. Mildred was a very pretty
girl, with fair pink and white complexion, and golden
brown hair. Her hat and habit were more stylish
than those of Eva, and she was mounted on a handsome
dark horse. Though differing in many respects, a true
friendship existed between them, and the two girls
greeted each other with genuine pleasure. Mildred had
the prettiest ringing laugh, that did one good to hear.
She was more of a general favourite than was Eva; she
had plenty to say for herself, and was never at a loss for
conversation. Everyone sang her praises; she was ever
spoken of as a sweet lady-like girl, and reigned queen
among her brothers' acquaintances, who would have
done anything to win a smile from her. Eva, on the
other hand, was just as shy and reserved, but her nature

was far the truer; she had not any amount of small talk ready, it is true, and she could not summon so many admirers to her side by her gay laugh and ready wit. But where Eva gave her love she never wavered or changed in her affections. With the pretty Mildred it was very different, she was more like a butterfly fluttering from flower to flower, gathering all the sweetness for herself, without giving a thought for the happiness of others, and when parted from her friends she soon forgot them, for she could make as many new ones as she wished.

" It is a full hour to sunset," exclaimed Mildred,—" do let us go through the wood home, it will be so charming ! "

No sooner said than they set off at a full canter, and presently found themselves at the entrance to the wood. Reining in their steeds to walking pace, the girls went leisurely down the narrow path, where in some parts rough stones were placed one above another almost like rugged steps. On one side was a steep declivity overgrown with ferns, heather, and other vegetation, interspersed with bushes of scrub. The sunbeams peered through the long vista of pines at the far end of the wood, framed in by the arching branches overhead, giving the appearance of a huge Gothic window, while the sombre stems and deep blue-green foliage of the trees were lit up with a rich red glow, making the wood appear indeed a temple of a thousand columns. Suddenly the two girls came upon a little group of children, busily engaged with some wild flowers, which they were tying in bunches. Whether it was the sight of the children, or the report of a gun which echoed up the glen, which startled Beauty, it would be hard to say, for he suddenly started aside, lost his footing, and plunged violently backward down the steep bank, bringing moss, stones, and turf with him. In an instant, with admirable self-possession, Eva let the reins go to their farthest extent, thus giving him every help in her power for righting himself. Happily the fall was broken by his coming in contact with the roots and brushwood, enabling him to stop in the descent after a severe struggle.

Eva, somewhat shaken, still maintained her seat. Mildred sat perfectly helpless, and screamed aloud. Presently there was a crackling among the dry leaves and twigs, and some one emerged from between the trees, followed by a large retriever dog. The frightened pony again began to rear, but Hubert—for it was he—sprang forward and caught the bridle.

He had heard Mildred's cry, and hurried to the spot. Both girls were glad enough to have Hu with them after their fright. Throwing his gun over his shoulder, and still holding the pony, he turned for home. Beauty was quiet enough now, and allowed himself to be led, and he hung his head as if he knew that the disaster was all his fault, and that if he got punished he deserved it.

" Eva," said her brother, " I wish you would be prevailed upon to give the pony up, you will ride him once too often one of these days ; he shied into a hedge the other day, and as to wagons, why, he seems to think they are on purpose for him to stick his heels in, and he turns into every gate he comes to. I must say I can't see what pleasure you find in riding such a brute."

" Nor I," said Mildred, who it must be confessed rather envied Eva's accomplishment.

" What can I do, Hu ?" said Eva sadly. " I have had

him ever since he was a colt, and I feel as if I could not part with him now."

The group of children who had rushed away alarmed during Eva's accident now returned with their hands full of lilies of the valley, which they begged the young ladies to buy. Mildred and Eva were delighted, and filled the pockets of their saddles as full as they would hold. They were very glad to get a glimpse of the small white gate which led to the road, and soon the friends had parted company, and Hubert and his sister, with panting, foaming Beauty, made the best of their way homewards.

CHAPTER V.

THE PICNIC.

CHAPTER V.

THE PICNIC.

IT was bright glowing summer; cherry and apple blossoms had disappeared, buds had formed into fruit, and soon it would hang rich and ruddy everywhere. On this particular June morning Meeta was up earlier than usual, for it was her birthday, and the Haverleighs were coming to spend it with her, and they were to have a regular picnic in the wood. Eva had made a cake especially for Meeta, frosted over with sugar; another loaf of cake also was browning beautifully on the stone hearth before the great log-fire, where the kettle was suspended by a chain in the wide chimney. By four o'clock in the afternoon, all arrangements were completed; cups and saucers, knives and forks, were all stowed away in the donkey-chaise. Alice soon harnessed Neddy into the shafts, and then there was nothing more to be done but to wait until the company arrived. The Haverleighs were a large family, consisting of nine children, some of whom had married and gone to live in another part of the country, but it is with the younger members that our story has to do. They lived in a fine old-fashioned house about four miles distant, and the two families had been intimate from childhood, the Rothesays at that time being tenants of Squire Haverleigh. Mrs. Haverleigh was exceedingly delicate, and ever since her children could remember had been an invalid, but she was always gentle and patient, and the quiet influence of her presence, as she lay day by day on her couch, was felt by all her children. Especially was it so with her boys, who looked on their mother with a feeling akin to reverence. Bertrand, the eldest son at home, often accompanied Hu in hunting or shooting; then there were his two sisters, Mildred and Leila, the former of whom has been before introduced to you, and two younger brothers, Ronald and James. Ronald was a midshipman in the navy, and it was chiefly owing to the tales he told of seafaring life that made Roger so anxious to go to sea. He never thought of the dangers and privations that have, alas! been so often experienced by thousands of our brave sailors, nor of those who had set out perhaps as full of life and hope and aspirations as he, but who had never come home again, and none knew what their fate had been. At this time Roger only saw the bright side of the picture his imagination had conjured up in such glowing colours,—all that he was to do and to be, how he should see the different customs and manners of foreign countries, which were so much more delightful actually to observe for himself than only to read about in books. Ronald Haverleigh was a thorough sailor; honest, true-hearted, and brave, he could brook the rough part of his career as well as the smooth, and carry it off so lightly in his free-and-easy way, that Roger, like many others, had no idea of what he really had to undergo in his profession. There was scarcely a ship afloat, from a fishing smack to a man-of-war, that had not been modelled and rigged, at one time or another, by Ronald's busy fingers. He would spin long yarns to his eager listeners when he came ashore, and he would tell them of the gorgeous sunsets, and of the size and distinctness of the stars, until many besides Roger wished that they could go to sea too.

But we are diverging from the picnic. By this time the rumbling of wheels was heard, and the clattering of hoofs in the distance; now it is more distinct, and the eager expectant little party run to the gate to catch a first glimpse of their friends as they turn the lane. Yes! there they are, there is the low pony-carriage, in which are seated Mildred and Leila, and there are Ronald and James riding on their ponies Snowberry and Jaspar on either side of the chaise, the gold lace and buttons of the young middy flashing in the sunlight. The two brothers are very unlike. Ronald is ruddy and sunburnt, with merry laughing eyes shaded by long dark lashes—he looks every inch a sailor; his thick chestnut curls cluster over his forehead, and his cap is thrust to the back of his head. Jamie is as delicate as a girl, and when you speak to him he raises his large dark eyes to yours, but both brothers have the same shining rings of hair. At length, the gate reached, they dismount, and there is much shaking of hands and hearty welcoming. The ponies are put in the stable, the girls crowd into the donkey-chaise, the boys preferring to walk, and away they go to the wood, Nettie bringing up the rear with the kettle. There are deep ruts in the field, and lump, bump, goes the little chaise over the grassy hillocks and through the hay. Driving through a hayfield is no easy matter, and every jolt occasioned fresh bursts of laughter, and little shrieks of alarm, and it seemed as if they must be precipitated on to the ground. But merry, hearty, happy children are not much troubled with nervousness.

The haymakers were busily at work and looked terribly hot; there were their basins that had held their midday meal tied up in handkerchiefs, and jars of beer stood here and there in the shade of the haycocks. The boys scrambled over a high gate stuck full of briars, instead of going round to the little wicket which led into the wood. Soon they found a hollow in the ground where there was plenty of peat and dry boughs that had been blown off during the rough winds and rains of winter. This hollow formed a capital shelter for making a fire and boiling the kettle. The boys collected a large heap of twigs, dry moss, and pine cones, and Roger, striking a match, soon kindled a famous fire, which blazed and crackled gloriously. The kettle was placed on the faggots, and Mrs. Rothesay and Eva began to prepare the tea.

They spread a cloth upon the grass, and then commenced unpacking the hamper. Eva cut the bread-and-butter, and there were plenty of cakes, marmalade, and strawberries and cream, and by the time that all was ready the water boiled, and everybody watched the process of tea-making with great interest. They then set to work and dragged the hay-cocks all round them, piling them up into great hills to protect them from the heat of the sun. Neddy was unharnessed, and allowed to roam about at will, but he did not behave very well, as he would persist in putting his nose into the plates of bread-and-butter, and Roger allowed him to drink some tea out of his saucer, which he seemed to relish very much indeed.

Tea at length being over, they began to play at different games. First they had " Hunt the Hare." The course the hare chose was into one of the covers, a pretty place where there was generally a nutting party in the autumn.

From the entrance there was a wide path completely overgrown with grass, extending some distance down the cover, bordered on either side with ash, silver birch, aspens, and nut-trees, in some parts so covered with lichen as to appear as if powdered with hoarfrost.

Very soon this path came to an end, and then there was a thorough scramble through brushwood and stinging nettles, which towered in some places above their heads. It was such hard work that after a while the girls, all except Meeta, gave it up, and made the best of their way back, often losing themselves in the attempt. As they had brought their bows and arrows with them, there was no lack of amusement. Eva, who had a very correct eye, hit the bull's eye several times, but some of the others did not succeed so well, the arrow glancing in every direction but the right. Some time after the boys appeared triumphant with the hare, who was very glad to have a rest on the hay. Then followed a game at cricket, but Jamie begged off, and was content to look on, leaning as he stood with his arm on the rough neck of Neddy as he was busily grazing. The last beam of the setting sun rested on his dark curling hair, and the boy looked like some old picture : his slight figure clothed in jacket and trousers, the wide collar and ribbon tied loosely about his neck, and above all his sweet and delicate features, made a *tout ensemble* that any artist might be well pleased to have studied. Twilight was deepening, and they began to make preparations for their return home. There was much laughing and fun in getting their things packed, and they were soon jogging homewards, the boys shouldering their bats and following after. When they reached the house they did ample justice to the home-made cake and wine set out for their reception; but increasing dusk warned the visitors that they must not delay, so with many expressions of friendship, and warm invitations to the Rothesays to come and see them, Mildred and Leila were soon ensconced in the pony carriage. The two boys mounted their steeds, and away they cantered down the field, followed by the eager eyes of the little company, until they were lost to sight amongst the trees at the turn of the lane.

Alice still remained when the others had returned to the house, and resting on the gate, watched the progress of the night as it crept on. What a hush seems to whisper in the dim perspective! There is no sound: the busy insect, the singing birds, and every animal on the farm have sunk to rest. As she walked slowly home through the gloaming, she watched the silver crescent of the moon rise high in heaven, while airy cloudlets floated in her wake, and myriad stars twinkled and shone in all directions. There was one in particular that little Meeta called her star, because it looked straight down on her, through her bedroom window, as she lay in her snug white bed at eventide, and every night the little girl fell peacefully asleep beneath the clear cold light of that ever-watchful star.

CHAPTER VI.

EASTERTIDE.

CHAPTER VI.

EASTERTIDE.

SOME months have passed away since the incidents recorded in the last chapter. Ronald Haverleigh had gone back to sea again some time ago, H.M.S. *Hero* having set out on a cruise, visiting several stations on her way, and he was not expected home for two years or more. Autumn and winter had come and gone, and now and then, as is often the case, the early spring would be accompanied by cold dreary weather, the winds would blow in cutting gusts from the northeast, as if winter were loth to say farewell. It was just such weather as I have been describing, when the sky was so overcast that you might expect snow at any moment and not be surprised to see a regular downfall, when Eva, well wrapped in a warm ulster, sallied forth one morning to look after and feed her numerous progeny. Rain or shine it was much the same to Eva; she did not trouble to look up at the clouds, or to tap the barometer, ere she made up her mind to encounter the chill, moist, uninviting atmosphere of the outer air. With quick steps she trod lightly over the thick drenched grass, that was quite of a blue-green tinge on the rising ground, while in the furrows and under the budding hedges it was brown and sodden enough. When she had satisfied the hunger of her ravenous brood, she set

off again, but not towards home just yet. The face of
nature even in her most sombre moments wore an
aspect that was pleasing to Eva ; she rather enjoyed
facing the elements, and battling with a stiff breeze.
The clouds were rising thick and gloomy from the grey
horizon, the trees in the distance reared bare and gaunt-
looking into the sky, and where the waters were out, as
was the case in some places, they looked like the masts
of sunken vessels,—so Eva thought, as she stood for
a moment gazing at them. Presently she fancied she
heard a faint twitter somewhere near at hand, and she
immediately began to look around, wondering whence
the sound could come. There was an old shed not far
off from where Eva was standing, and she hastened up
to it, thinking some bird might have fallen from its
shelter in the thatch numbed by the cold, and she was
not altogether wrong in her surmise, for she soon heard
the chirp again a little louder, but still very feeble.
Eva was on the alert now, and nothing would have in-
duced her to return home without her bird, whatever it
might be, and she began to tread slowly and cautiously.
At last, nearly hidden in the grass, and lying close to
the shed, she discovered the object of her search. Eva
took it up and covered it with her hands for warmth,
and as she inspected her new pet she found it was a
sparrow,—only a sparrow ! one of a large and numerous tribe, a saucy, brown, pert little bird that can be
seen in numbers anywhere and everywhere; but Eva
did not despise her newly-found treasure for all that,
her only trouble was that she feared it was too old to
rear, for it had evidently fallen from the nest in its
haste to try the wings which at present were too weak
to bear its weight, and if it had not been discovered, it

must have succumbed to the cold and damp. It soon revived, however, in Eva's warm hands, and when she partially removed her fingers to see how it was progressing, two bright little eyes looked up eagerly into hers.

As soon as Eva returned home, it was the work of a very few minutes to find a box and place the bird in it with some wadding. Then off she went again, and soon reappeared with a cup of soaked bread and bruised seed. The little foundling was very troublesome at first, and would not understand that his kind benefactor wished to save his life, which could only be done by forcing food down his throat, as he resolutely refused to open his beak to receive it. But this only continued for a time ; as he grew stronger he became ravenous, and Eva fed him with crumbs on the point of a pin, while the bird flew round in eddying circles, fluttering his wings all the time. Though merely a fledgling, for he could only have chipped the shell a few weeks at farthest, he showed himself a very wise little bird notwithstanding. He did not trouble his head by thinking of his past sorrows—doubtless in his small bird's memory they were already forgotten—he only knew in some way or other he had become possessed of a delightful bed and plenty of food, and that was quite enough for him. As time went on, the sparrow grew and became a favourite with the inmates of Fern Glen, taking his place a recognized member of the household. One day Eva might be seen busily scrubbing an old cage, that had been dragged to the light to meet the growing requirements of her young favourite, which was daily becoming more amusing. While Eva was dressing in the morning, he flew after her wherever she went, alighting on her head or shoulder, often greatly hindering her progress by creeping up the sleeve of her dress, which he was very fond of doing. She now supplied him with food, as he was quite old enough to feed himself, but he did not like the trouble, and flew about with the pin in his beak with which Eva used to feed him as a mute reminder.

As the days grew longer Eva much wished to make use of the bright mornings to get an hour or two at her painting or sketching before breakfast, and was wondering how she could manage to awake at an early hour, when all at 'once she thought of her sparrow, and whether he would answer her purpose. Accordingly, after she had fed him, which she always did last thing before going to bed, she left the cage door open and retired to rest, anxious to know if her project would succeed.

The next morning, when the rays of the sun had just looked in at the window, Eva was awakened by a gentle flutter of wings over her head, and when she raised her hand, the dear little fellow nestled into it, popping out his saucy little head, seemingly delighted that he had found his mistress. He was evidently very fond of the human race, and never seemed to wish for feathered company, nor had he any desire to fly away, for one lovely afternoon little Meeta opened the window, forgetting all about the bird, which immediately flew out into the garden. Eva and Meeta looked on with great consternation, but almost before they could exclaim, back he came again, alighting on Eva's neck and hiding in her hair. For not the loveliest day, with its wealth of sunshine and many allurements, could tempt him from her who had rescued and cherished him.

All this time spring was advancing, and at length Eastertide had come round once more. The season was somewhat tardy, but when it arrived it came with smiling face and genial atmosphere. Leila Haverleigh had come home again for her holidays, and was very soon busily engaged in church decorations, in which she always took great interest, gathering the freshest and best flowers her little garden afforded for the occasion. And very sweet the church looked on Easter Sunday, as the spring sunshine gleamed on the wreaths of azaleas and camellias, interspersed with ferns, laurustinus, and other evergreens. Lordly arum lilies reared their silver cups and graceful leaves on either side of the chancel, and the delicate lily of the valley embedded in moss clothed the font. Leila had great taste for flowers, and took a pride in her garden, which was situated in one corner of the grounds, divided from the road by high iron rails. In the spring time she had primroses, violets, hyacinths, and red and white daisies, and later on in the season there were stocks, pinks, and verbenas, so that amidst the labyrinth of greenery Leila's little garden had a very pretty effect. Knowing her love for flowers, a friend had given her some gladioli roots, and for a long while the little girl had watched them, from the time when the delicate green shoots had just shown above the ground, until they had grown tall handsome plants, with slender reed-like leaves. And soon on each had appeared buds of varied tints, which promised in time to become beautiful blossoms.

One morning Leila came running into Mildred's room just as the latter was descending to breakfast. "What is the matter, Leila?" she asked, for Leila had burst into a fit of crying, which was most unusual with her.

4

"Oh! my beautiful buds," she exclaimed, "that I have been watching for ever so long, and I was keeping them for poor mamma to put in her room, for she thinks them so pretty, and they are all gone!" she said with quivering lips,—"not one left."

"Who has touched them?" said Mildred indignantly. "I will tell Bertrand, and the person shall be well punished.

"It was not a person," said Leila, "but some naughty little children. I was just going down to my garden this morning, expecting to find some of my buds opened, and do you know, as I came in sight of them, I was just in time to see one of the children putting her hand through the rails and gathering the last bud. It was not that she wanted them, but just for mischief, for she threw them all away. I dared not speak to her, for she was as impertinent as could be."

"What a shame!" said Mildred. "I will go and tell my father. I daresay they will come into the garden next and gather the roses. I do dislike the children of the present day, they are so rude, and no one seems to know whose business it is to correct them."

Squire Haverleigh laid down his newspaper as the girls entered the room. He saw in a moment that something was amiss.

"What is the matter with my little Leila?" he asked, as she came up to him.

"Oh! father," she said, "all my gladioli flowers are gathered, not one left!" and the tears sprang to her eyes again, and Mildred finished the story.

"I call this too bad," said the Squire; "the fact is, if there were fewer school-treats and a little more wholesome discipline, I think we should find a vast improvement. I have noticed for some time the growing impertinence of the children about here, and it has caused me much uneasiness."

"Only think," said Mildred, "they gathered the buds and threw them away, while Leila was standing by; if they can do so before one's face, there is no knowing how far they will go when one is absent. The people here make so much fuss about the religious training of the children, and yet what a noise and talking they make when coming into church! they are much better behaved at Fern Glen; Mr. Cecil, the rector, says there is no religion without reverence, and I believe him."

Bertrand, who had been listening, said with a smile, "How angry you were, Mildred, with the gentleman who had a girl taken into custody for plucking a flower! it strikes me this is a similar case, except that his might have been a much more valuable specimen than Leila's gladioli."

Mildred coloured. "I know I was," she said, "but it sounded so absurd in the papers. I feel for him now, for I daresay he was as annoyed as Leila, provided he had been watching his flowers day by day, as she has done."

"Let this be a lesson to you, my daughter," replied her father, "never to judge another until you have heard both sides of a story. I would buy Leila some more directly," he continued, "but if I did they would not give her half the pleasure her own would have done; however, I will have some wire-work put this very day to protect her garden from any more such innovations," and the Squire's kind interest helped Leila in a measure to bear her vexation.

CHAPTER VII.

THE COLLEGE FÊTE.

CHAPTER VII.

THE COLLEGE FÊTE.

AS time went on, the great event of the year was drawing nigh, which was a grand *fête* that was held annually at Roger's school, and it was usually arranged that there should be a flowershow on the same day.

Long before the holiday took place, it was the topic of much conversation and excitement amongst the young people of the quiet village. Leila Haverleigh, with some of her young friends, were going to compete for prizes for wild flowers, so that Leila had plenty to take up her attention just now. The families of all the boys who went to the college were provided with tickets, and were expected to be present.

When the day arrived, numbers of people might be seen making their way to the fine old grounds, for there were to be walking-races, running-races, foot-ball, and gymnastics, and to the successful competitors prizes were to be given. On this eventful morning many carriages of various descriptions could be seen rolling along, " all on pleasure bent." The college was a handsome building of red brick and stone facings, studded with numerous windows, and a colonnade consisting of stone archways stretched the whole length of it. The field in the foreground, in which the sports were to be held, was surrounded by rustic palings, and groups of village boys and girls stood peering in, taking a lively interest in the proceedings. A crimson cord enclosed the space set apart for the races, and blue flags marked the route for the runners. Bertrand Haverleigh, who was mounted on horseback, kept a sharp lookout on all intruders who ventured within the enclosure. It also devolved on him to ride round during each race to see fair play. At length the boys, equipped in caps and suits of various colours, began to congregate as the company arrived. Presently a low phæton appeared in sight, in which was seated a lady in deep mourning, accompanied by her two daughters and pretty fair-haired little son, her youngest child. While the preparations are going forward, we shall have a few moments to give you a description of her. "Lovely Mrs. Arlington," as she was still called, was a widow, and she, with her numerous family, resided in a solitary heavy-looking stone house, some little distance from the town. For the sake of her sons she had come to live near the college, that they might receive the advantage of a classic education. The boys were classmates of Roger's, being in the upper school, and one of her sons, who was in the navy, was a friend of Ronald Haverleigh's, so that there was an intimacy existing between the three families.

At length came Mrs. Rothesay's pony-carriage with Eva and Alice. Meeta was mounted on Neddy, and trotted along by the side. Hubert was already there, looking out for his brother Roger. The girls were quite interested in deciding who would be the winners as they scanned one and another of the groups on the field ; and were not a little amused when a wagonnette and pair drove up, and a youth, very gaily dressed in bright blue and white, descended. It was apparent that the young fellow thought quite enough of himself.

"Do look at that boy, Hu," said Eva; "I do not believe he will win a single race."

"Why," said Hu, "what is the matter with him?"

"He is not the make for it," replied Eva, "he is such a heavy build."

Just at this moment young Harold Arlington appeared from amongst a number of boys who were standing near, equipped in a white suit with no colour at all, but it set off his lithe well-made figure to the best advantage.

"There," said Eva; "now, Hu, can you see the difference? I would back Harold against all those I have seen in the field to-day."

"Oh, everybody knows that he is a first-rate runner," replied Hu, "but he is to be this time placed with a number of bigger fellows."

"That is too bad," said Eva; "I shall be so sorry if he does not win a prize."

Presently Mildred, Leila, and James arrived, and drew up their carriage by the side of Mrs. Rothesay's, and they were soon joined by Hermie and Linda Arlington, who were discussing the points of the different competitors.

"Mother is sorry that our boys are going to run," said Hermie; "she is so afraid of their over-exerting themselves, but they will have their way. Let us go and see the prizes," she continued, "some are so handsome."

On one side of the field there were tables covered with a crimson cloth, on which were displayed silver cups of different sizes, some of which were richly chased.

There were also some silver sculls for the winners of the rowing race, which was to take place the next day, when there would also be a grand cricket-match amongst the senior lads of the college. At length the company within the ring began to wear a business aspect, as the proceedings of the day were about to commence. On a raised platform, surrounded by banners and different devices, was seated a brass band, which soon began to pour forth harmonious strains. The sunbeams glittered on the instruments, on the bright silver prizes, and on the wheels of the numerous vehicles as they rolled past.

Coloured favours fluttered in the breeze; crowds of children stood against the crimson cords, clad in bright summer attire; seats were filled with visitors; and altogether it was a gay and pretty scene.

First there was a walking race of three; one was rather a lean, cadaverous-looking lad, of the name of Bennet, who looked as if he had been in training for some time. The first round they kept pretty well together, but after a time the one next to Bennet began to fall off slightly, giving a hop and a spring every now and then to recover lost ground; but it was of no use, it was plain to see that steady-going Bennet, who had not altered his pace for a second, was decidedly gaining on the other two every moment, though he began to pant somewhat as he neared the goal. Presently there was a shout of " Bennet has won ! " which was followed by much clapping.

A hurdle race followed, in which a dozen or more competed, and as Harold Arlington was one, a good many besides his friends looked on with eager interest, as he was a general favourite. At a given signal all started off, some running as if for their lives; but

Harold was quite cool and collected, and everybody began to get breathless as the boy allowed several to get ahead of him. Now he began to quicken his pace. Hurdle after hurdle was leaped, and Harold was evidently gaining ground on those before him. Some could not help smiling at, as well as pitying, one poor fellow in red and white, who kept patiently on, though he was what schoolboys would describe as " nowhere." His body was very long, and his legs very short, and he gave one the impression that it was as much as he could do to carry himself. At this moment someone shouted out, " Arlington will win, look at him !" and there was Harold, first and foremost of the throng. There was no doubt now who would have the prize, though it was a hard-fought contest, and everybody was glad when he was caught safely in the arms of the judges at the goal.

One of the lads had to be taken behind the scenes and have a cordial administered to bring him round, for many of these races were terribly hard work. After the hurdle race, there were shorter distances, and Harold, who seemed none the worse when he had rested a little, seated himself on the ground, and taking off shoes and stockings, ran the remaining races, in which he competed without them. He still bore the palm, being successful in each. Then followed the high jump, the pole jump, and throwing the ball, after which was a race in sacks, which created a great deal of amusement. Lastly, to finish the afternoon, a number of quite little fellows ran. The distance was long for such small boys, besides which they had to take a large ball round with them. One little fair-haired boy had the advantage, keeping the ball before him all the time, and his spirits were kept up by the bigger fellows as they shouted out, "That's right, keep up, Charlie; go it, Charlie! you'll win," and he did too, and reached the goal, amidst a tremendous cheering.

Eva was right with regard to the dandy in bright blue and white; he ran two races, but did not win either.

The company now repaired to the flowershow, which was held in tents at the other end of the field. There were some very lovely specimens of hot-house flowers and ferns, and one part was devoted to fruit and vegetables.

Soon little Meeta came running with glistening eyes to her mother, to tell her that Linda had gained the first prize for wild flowers and Leila the second. Meeta was a generous child, and was as pleased as if the prize had been awarded to herself, for she had been one of the competitors, and had taken great pains with the arrangement of her flowers; indeed, they were so prettily interspersed with moss and grasses, that it was generally thought that Meeta would have had one; they were commended, but could not be classified. Eva had gained the first prize for a very pretty design in butter. The show of fruit was particularly fine, some of the strawberries being of enormous size. Tea and coffee were served in one of the tents, and the company were very glad to partake of some refreshment after the fatigues of the day. The college lads, as well as the rest, did ample justice to the rolls and cake. When the band struck up again, many of the children began to dance, and Squire Haverleigh, who was looking on, remarked that he thought it was the prettiest sight he had seen that day. As it grew dusk there were preparations for the fireworks, which were to conclude the day's festivities. The Squire mounted Meeta and Leila on one of the seats in front, where they could see beautifully. He was the kindest man, always looking out for those who kept in the background, and by his easy cheerful manner contrived to put the most timid at their ease. He was particularly fond of Alice, whose modest demeanour pleased him. The display of fireworks was very fine, reminding some of the company of the gorgeous exhibitions at the Crystal Palace, only the successive hues of brilliant rose and violet, green and white, rested on the trees and surrounding country, instead of on the snow-white feathery spray of the fountains as it falls into the lakes beneath.

At length the various equipages began to move from the field, and everybody after a warm leave-taking was soon turning homewards. The inhabitants of Fern Glen were an united people, kind and warm-hearted, and there was as little cliquism amongst those who assembled together that day as could be found anywhere. Perhaps that was the reason why everybody went away with such a kindly feeling towards all whom they had met, and why they were so satisfied with the day's entertainment.

CHAPTER VIII.

A RIDE TO WESTON UNDERWOOD.

CHAPTER VIII.

A RIDE TO WESTON UNDERWOOD.

MORNING lessons were over, and little Meeta had run off to feed her rabbits, while Alice seated herself on a low chair by the window, and prepared for a pleasant hour with her favourite studies.

Presently Eva entered the room. "It is such a fine morning," she began, "just the day for a ride. I will go over to Squire Haverleigh's and ask Mildred if she would like to accompany me. It is already so late in the year that if we wait much longer the leaves will be off the trees, and I had so wished to see the lovely autumnal tints before they are faded."

"I am glad you thought of it," replied Alice, looking up. "Mildred has often said how she would like to go with you, as you know all the prettiest spots; besides, Leila will be returning to school shortly."

So Eva hastened away to put on her habit, and was soon trotting her pony at a leisurely pace across the fields towards the park.

Some time afterwards three fair equestrians might be seen coming down the lane, bent upon having an afternoon's enjoyment. Over the rustic bridge they went, which echoed back the thud of each cantering hoof with a hollow sound—past the village and the town where Eva so often went on marketing expeditions, then on to a long road, which stretched far before them

5

in numerous curves and turnings. It was skirted by a low stone wall, by the side of which were long lines of trees extending as far as eye could reach. Just as they arrived at a place where two roads met, they were confronted by a hunting party in full cry, for they had just found.

Dashing along the road they came, sending the mud flying in all directions; the three girls had much ado to keep their ponies from following. Beauty lay back his ears and began to kick in earnest, but Eva, who was used to his tricks, held him in. On they came—the panting hounds running far before, over the low stone wall which skirted the road with a leap and a spring! It was a pretty sight, the gentlemen in pink forming a glowing contrast to the bronze tints of the surrounding country.

The girls remained until the shrill sound of the horn and the distant echo of horses' hoofs could no longer be heard.

"There," said Mildred, in a discontented tone, "that is just what I should like, but Bertrand is so ill-natured he never will take me. I wish Roy were the eldest, he always does whatever I ask."

Eva declared she did not care for hunting, it seemed so ridiculous, as well as cruel, to make such a tremendous fuss and noise over one poor little helpless animal. She for her part enjoyed a good gallop across the country for its own sake, and Leila agreed.

Mildred said that was because she was chicken-hearted, and dared not face a fence, but Leila was a good little girl, and only laughed instead of getting into a temper, for she knew it took two to make a quarrel, and it may have been the secret knowledge that she was a better horsewoman than her sister, for Leila had often been praised for her graceful riding; she was a better figure, too, than the pretty Mildred, and that was always a sore point with the latter.

Very soon they came to the lodge gates, surmounted by a high stone archway, where, on either side of the wide gravelled road, was a plot of grass railed in by chains. Their way led now through the archway on to a stone bridge, and in its pillars were niches, reflecting themselves in the clear waters of the Ouse below, where children loved to sit when they came violeting in the spring-time.

Just at this point the water was broad and deep, and white swans could be seen pluming themselves on the bank in the distance.

Leaving the bridge they cantered down the undulating field, taking care to keep clear of the old trees which grew about their path, stretching their low overhanging branches far and wide. On their right hand was the rustic church, situated at the edge of a wood, with its picturesque churchyard surrounded by evergreens and enclosed by iron rails.

The entire place was a bed of moss, where violets and primroses grew in rich profusion in the spring. It was a spot so sacred and retired that Gray might have chosen it in which to pen his " Elegy."

After passing through several gates, which Eva opened dexterously with her whip-handle, they came to the water's edge, where each had to go in single file, and far away the tangled spinnies rose like islands in the river as it went meandering onwards and vanished in the hazy distance. At length they reached the village outskirts where once a poet dwelt secluded, whose memory still is fondly cherished for his songs of tender pathos, penned amidst the calm of nature, broken only by the warblings of the nightingales and thrushes, and the numerous feathered songsters. They were his only auditors, permitted to be the recipients of his deep thoughts, that rose upwards and found utterance in the sweet verses so well known and well beloved—for the poet Cowper will never lose his place in English hearts, and is still read by the fireside when evening closes in and all is bright and warm within the house.

I doubt not that many who read this story have heard of "the poet's cat sedate and grave," and of "The Dog and the Water Lily;" and who has not read of the three hares, Puss, Tiney, and Bess, which would gambol round him and cheer him with their playful ways when his heart was burdened and sad?

Those pretty verses, too, commencing—

" The rose had been washed, just washed in a shower,"

would never have seen the light but for the timely interference of an intimate friend of the poet's, for in one of his desponding moods he was about to consign it with other manuscripts to the flames. These verses, in their original delicate transcription, are still carefully preserved in the family of Cowper's friend.

At length our three equestrians were riding leisurely through the park-like fields, studded with clumps of elms. On their left was the wilderness, with its white monuments erected to favourite animals gleaming here and there from among the trees, one bearing the well-known epitaph * to the famous pointer belonging to Sir John Throckmorton.

* " And to all this fame he rose
By only following his nose."

Before them in the distance was the lordly avenue of limes; though to see them in perfection was in the early summer, when the sunbeams would shine in amongst the transparent and delicate greenery, and the air would be filled with the fragrance of the lime blossoms. This avenue led to the alcove on the hill described by Cowper in his poems.

"Why," said Eva, consulting her watch, "it is not so late as I thought, we shall have time to call at the Lodge—I have been so often invited. What do you say?" she added, turning to her companions. Both girls exclaimed at once that they would like it of all things, for Eva's friends occupied the house which had once been the home of the poet, and it was with no little curiosity as well as pleasure that Mildred and Leila looked forward to an introduction.

Soon the three girls were trotting their ponies up the quiet village, and presently halted before the house. It was a substantial country residence, studded with two lines of windows; the lower ones nearly reached the trim narrow walk which ran round a pretty flower bed on either side of the door. Iron rails divided the fore-court from the high pebbled path outside, and six steps led down into the road. Stone flags paved the way to the front door, which was painted dark green and embellished by a bright and ponderous brass knocker.

Eva dismounted, and presently the door was opened by a maid-servant, but before Eva had time to speak, Miss Lucy, the sister of Mr. Ashwell, hastened out and begged them to come in. "I am so sorry," she continued; "my brother is out hunting, but I expect him in every minute."

At her bidding the girls rode round to the stables,

giving their ponies in charge of a groom who stood ready to receive them. Very soon they were walking through the cheerful-looking hall, with its pretty black and white tiles, and its old-fashioned table, above which hung a stag's head.

As they went upstairs, on the right hand side of the landing was the spare room, with its two wide pleasant windows, and Miss Ashwell pointed out the lines on the shutter of one of them written by the poet on his departure from the peaceful village—

"Farewell, dear scenes ! for ever closed to me,
  Oh for what sorrows must I now exchange ye !"

How many years have they remained to bear their mute record, treasured by loving hearts, for no careless hand has been allowed to obliterate the faint characters made sacred by the seal of death !

The girls then descended to the dining-room, with its brightly-panelled walls ornamented with pictures, and very glad they were to rest after their long ride.

Presently there was a loud knock at the door, and Mr. Ashwell entered. Very kindly he welcomed the visitors, begging them most warmly to take tea, which should be forthcoming almost directly. It was of no use their making any excuses, because he would not hear them.

Mr. and Miss Ashwell were known for their kindness and hospitality to all who called upon them, everybody, whether strangers or friends, receiving a hearty welcome.

He told the girls how sorry he was that he could not accompany them part of the way home, but the fact was his horse was at that moment reeking in the stable, too tired to go any farther, as they had had a long run.

While the maid was bringing in the tea, Miss Lucy brought out a book, wherein were written the autographs of many a notable personage who had made a pilgrimage to the residence of the poet. When the meal was finished, she took her guests round the garden and orchard, and there was so much to interest that Mildred and Leila would have forgotten how time was getting on, had not Eva reminded them of the long distance before them, and that it would be dusk ere they reached home.

Their kind friends both came to the gate to see them off, Miss Lucy anxiously enquiring if the animals were safe ? " Oh," said Eva, " mine is as quiet as a lamb when he is going home." But it seemed as if the stout sturdy pony resented such an epithet applied to him, for directly Eva attempted to mount he began to rear.

Miss Lucy, who was alarmed, quickly retreated, saying she did not call that " as quiet as a lamb." " He has made too free with the corn," exclaimed Eva, laughing.

The girls now started off at a brisk pace down the road, for clouds began to gather, and the prospect of a coming shower warned them that they must not linger. Once in the vale below, they saw a solemn owl with buff plumage rise and sail away to the sombre woods beyond. Presently Eva remarked that she thought with all their haste the rain would be before them, adding that she knew of a short cut, but it was rather awkward, as the path was narrow and rugged. Soon they came to a gate ; the rusty latch was difficult to unfasten, but this obstacle overcome, they reached the path mentioned by Eva, a rough place strewn with large loose stones. One false step would have precipitated them down the steep bank into a dark pool of stagnant water covered with duck-weed, anything but a pleasant place in which to alight. They were glad when they had safely passed it and found themselves in the high-road again.

In a few minutes they espied Bertrand Haverleigh some distance before, who was returning from the hunt.

Mildred called to him, and wheeling his horse round, he came and joined them. "You must make haste," he said, "there is rain in the wind, and the best thing you can do is to make for the shed yonder." Opening the gate for them, each passed through, just as some heavy drops began to fall.

"Now for it," said Bertrand : "who will be there first ? " and then they gave the ponies their heads and made for the shed, helter skelter. The rain now beat in their faces in earnest. Leila's pony seemed as if he were flying; she soon lost her hat, and her long hair was tossing in the wind. Bertrand kept in her wake, hooking up the hat with his whip as he rode along. They were glad enough when they reached the shed, and could stop to recover breath. The three reeking ponies snuffed about to see if they could find anything to eat in the manger, while Mildred and Eva endeavoured to put their dishevelled hair a little straight before they started for home.

At length it began to grow lighter, and the four set off again down the field. When they came in sight of the house Bertrand reined in by Eva's side. "You will come in ? " he asked ; "my mother will be pleased to see you." Eva thanked him, but explained that they would be watching for her at home. So bidding her friends "Good-bye," she started off at a brisk canter towards Fern Glen.

CHAPTER IX.

THE RECTOR AND HIS PEOPLE.

CHAPTER IX.

THE Rev. Theodore Cecil, the rector who had recently come to Fern Glen, was opposite in every respect to his predecessor, who was as proud and taciturn as Mr. Cecil was kind and affable. Already he had effected changes that had proved most beneficial to his church, and his coming was regarded by the inhabitants with sanguine expectations. Mr. Cecil could not be styled a learned man; he was not an orator, nor did his sermons rise to eloquence; but they were plain, practical, and thoroughly comprehensible, sometimes imbued with a vein of poetic feeling, caught from the contemplation of the peaceful beauties of rural scenes by which he was surrounded. It was his large-heartedness, his ready sympathy, his love of children, and his friendly and agreeable manners, which had won for the young rector the affections of his people. Moreover, no spirit of wealth-worship had ever crept with subtle influence within the calm seclusion of the rectory. His family consisted of his wife and two children, about the age of Meeta, and very delighted was the little girl when at length they became acquainted, for hitherto she had never had any young companions of her own age. The little Meeta was a favourite with Mr. Cecil; she was one of the best-behaved amongst the children of his con-

gregation, and her bright intelligent eyes were generally
intent on his when he addressed the children especially,
as was often the case. Mr. Cecil was one who believed
it to be quite possible for little ones to sit still during
Divine service; he was not at all for taking the part of
the mothers when they argued that the poor little things
got restless, and that it was a long time to sit, etc., but
he steadily maintained that there is always something in
the service or sermon that even a child can understand,
and that no child ought to be allowed to be a nuisance
and a distraction to other members of the congregation
who had come for a season of worship and meditation.
From their earliest years every child in Fern Glen
was taught to look on the grey old church as the House
of God, nor would they ever dream of talking aloud or
whispering in the sacred edifice, as some, I am grieved
to say, do; but no rebuke was needed by little Meeta.
She would enter the church in as quiet and orderly
a manner as any there; she never made remarks on
the dress worn by other children, nor did she ever
look proudly on those who were not so well clad as
herself; as long as they had kind and pretty manners,
she would be friends with them at once, for being a
timid child she shrank from anything rude or boisterous.
Mr. Cecil disapproved also of talking immediately after
the service, and some thought he made too much of
trifles, and was a little too particular about such things;
nevertheless, he was much liked in spite of his plain
speaking, and there was not a more decorous congregation anywhere to be seen than at Fern Glen, though its
primitive style would be looked upon slightingly perhaps by more fashionable assemblies.

Little Meeta's pet animals were a constant source of

amusement to the rector's children, who were frequent visitors at the farm ; her example too was good for them, for she was always kind and gentle to all dumb creatures, never teasing or tormenting them, but treating them as her friends and playfellows, and that is why they loved her so.

I should deem it very improbable that you ever saw on any other farm such a perfect understanding as existed between the animals and inmates of Fern Glen. If you watched either Eva or Alice as they went across the fields, you would notice that the horses there would raise their heads, prick up their ears, and in another moment would be coming at a brisk trot down the grassy slope to meet them.

There is Bessy and her two colts ; the latter advance with a prance and a caper, while the former, a quiet sedate little creature, takes it more leisurely, for she is now on the shady side of pony life, and her playful days are over.

The colts are as roguish as kittens, and seize hold of Eva's jacket with their teeth, and are as saucy and impertinent as they can well be. But the girls only laugh and caress them, they are quite used to their language, which plainly says that they are ready for a game on the shortest notice.

But it is with Thisbe especially that I wish you to make acquaintance. There she stands, with the stout strong-looking cob, under the trees at the far end of the field.

Presently she catches sight of Alice, and uttering a low neigh, she sets off at once towards her, shaking the long mane from her eyes as she goes. She is in very good condition, and her coat shines like satin in the

sunlight. But if you had only known her a few months
ago, I am sure you never would have believed that
Thisbe and the wretched ill-looking creature, that then
made its first appearance at the farm, could be one and
the same. " What a miserable pony ! " you would have
exclaimed; but Mr. Rothesay knew too much of the
animal tribe to be deceived by a hanging look and
starved appearance. And so poor little Thisbe, happily
for her, changed masters. The days of drawing great
cart-loads, far too heavy for her, were now at an end.
Rough words and cruel blows she never will experience
again.

But you must not fancy that she changed for the
better all at once—oh dear no ! When taken by surprise
the frightened expression would return; she could not
understand the kind words and gentle caresses of her
mistress Alice, but would lay back her ears and look
quite wild. But Alice well knew that it was not per-
versity or a bad disposition, and continued coaxing and
fondling her, until she gradually became used to this
new and wonderful change in her circumstances. When
Thisbe became the property of the kind and humane
owner of Fern Glen, she was not only in miserable condition, but ill broken in, and he at once took her in
hand, and now, at the time of which we are writing, at
a single word or signal she would follow him anywhere.
Signs of a good breed too had become apparent in the
arch of her neck and the fire of her eye, and Thisbe had
grown a somewhat spirited, petted, handsome creature,
that was worth a good round sum of money. But she
is Alice's pony now, and I don't suppose she would have
parted from her pet for her weight in gold.

The girls were very fond of making sketches of their

favourite animals, and on one bright sunshiny afternoon
you might have seen quite a procession taking their way
to the barn. First, there is Alice with her pony, which
does not seem to appreciate the honour, for she looks
rather frightened ; then little Meeta struggling along
with two chairs. Last of all comes Eva, with her easel
and the etceteras attending it. Thisbe cannot understand it at all, and prances about in fine fashion ; but
after the first sitting was over she began to like it very
much indeed, and to look out for these pleasant after-
noons. For she had found in one corner of the barn
that there was a splendid load of dried clover and peas,
with which she soon became intimately acquainted.
She had put her nose, too, rather timidly to the easel,
but had found it quite harmless, and at length stood as
still as any animal would do under the circumstances.
After the sitting was over, she would follow Alice round
the barn, pick up her handkerchief when she dropped
it, and was fast becoming as intelligent as any pony you
may have seen in a village circus. Very frequently Mrs.
Rothesay would bring her knitting and join them, now
and then holding the goats or donkey in position for
the busy artists, and Neddy would rest his head on her
lap, blinking lazily as he stood. If the painting happened to be of any animal belonging to Hubert, he too
would put in an appearance, and stretching himself on
the hay with a book, take a critical survey from time to
time of the progress of the painting.

As the sun nears the west, it is the signal for the
little party in the barn to be putting up and returning
homewards. How sweet is the landscape at this peaceful evening hour ! The reflections of the clouds pass
rapidly over the sunlit grass, now the distant village is

bathed in his beams, now resting in transient shadow—
the mild wind plays airily with the grass tops, passing
ever and anon amongst them, until they appear like
white crested wavelets laving the feet of the little company as they walk rapidly along.

After tea Alice and Eva with their little sister sallied
forth into the fields, as was their daily practice, to look
after the refractory poultry which had strayed away, and
had not returned.

There was Pip, the peacock, who by his propensity to
roost out would greatly disturb his owners, as they
knew of dangers abroad in the shape of a wily fox, that
had several times pounced upon unsuspecting and foolish
fowls that preferred the wide range of the fields to the
safe shelter of their own snug beds at home.

It was one of those lovely evenings, when twilight is
loth to depart, and it lingered so long that the girls
lengthened their walk, and getting over several stiles
took their way across a wheat field, making quite a
round ere they would again come in sight of the house.

Eva and Alice were some distance in advance of
Meeta, who generally lingered behind, searching for
something new to take home with her. Suddenly, as
she neared the overhanging trees of wood and cover,
her footsteps disturbed something from amongst them,
and with a rustling and a flapping sound out flew a
bird and sailed away, uttering a strange harsh cry as it
went. The noise startled the still air, and it seemed to
Meeta as if the night herself were holding her breath to
listen, expectant that something more was to come! and
the frightened child, not daring to stay a moment longer,
rushed down the field as fast as her little feet could
carry her.

"Why, Meeta," said Alice, "what is the matter?" as the child reached her side, looking so pale that both sisters were alarmed.

"Oh!" said Meeta, when she could speak, "there was such a fearful noise in the cover yonder, and something black came out of it."

"Why, you foolish little child!" replied Eva laughing, "it was nothing but a bird."

"Never mind, May," said Alice, in her cheerful reassuring voice. "I have been startled myself sometimes at those birds, for they look so large in the dusk, but it is not as if you were all alone, you are quite safe with us."

Alice was interrupted by a gruff voice, which sounded close to them—

"Safe, did you say?—not so safe as you think."

The girls started, and looked at each other with awestruck faces, and turning, were just in time to see the figure of a man slouching off under cover of the hedge, and disappear amongst the trees near at hand. They did not know what to make of words that sounded so like a threat, and the incident formed the subject of discussion when they arrived home that evening. But Mrs. Rothesay wished to make light of it for her little Meeta's sake, and as the days went on, and nothing more transpired, the subject was seldom referred to, and appeared at length forgotten.

6

CHAPTER X.

THREATENING CLOUDS.

CHAPTER X.

THREATENING CLOUDS.

SUMMER once more held her sway over the small world of Fern Glen and its surroundings. In this particular year the season was unusually sultry; the sun, from early morning until the time of his setting, glared down on the burnt patches of grass, and on the drooping flowers; no cloud floated in the deep blue sky, dazzling in its intensity, and not a breeze disturbed the trees, which stood in the golden light motionless as in a picture. Harvest was drawing nigh, and the difficulty of procuring labourers was daily increasing, as everywhere their services were required. A spirit of discontent had for some time shown itself amongst some of the men in Mr. Rothesay's employ. One in particular was a man of the name of Giles Winter, who was dark and savage in aspect, of a morose temper and sullen disposition. He was continually grumbling about what he called "long hours and poor pay." He rented one of the smaller cottages on the farm, owned by his master, who was known to all as a just and a kind landlord. Mr. Rothesay did not like the man, but as he was a good workman, and very poor, he felt loth to turn him off, for he was pretty sure it would go hardly with him as to the obtaining of employment elsewhere. Winter himself knew very well that he bore a bad character

in that part of the neighbourhood; he was suspected
of poaching, but hitherto he had escaped the vigilance
of both keeper and police, knowing every nook and turn
of the woods and covers for miles round. He was
smooth enough when the master was by, though he had
plenty of evil remarks to make when he was not there
to hear. As has truly been said, "One black sheep
infects the flock," so by his continued complaining he
got others to join with him in his expressions of dis-
satisfaction; and so it came to pass that, just at the
busiest time of the year, when help was most needed,
a good many of the men refused to work unless Mr.
Rothesay acceded to their terms. But bright sanguine
Hubert, who always had a pleasant word, and was ever
ready to put a brave face on things, kept up his father's
spirits, and declared that he "would tackle the men,
and find out the grievance," and if the worst came to the
worst, why, they must just put their shoulders to the
wheel and help themselves.

But very soon after affairs began to wear a more
serious aspect. Threatening letters were often received,
and one morning, when the family came down to breakfast, many of the windows at the back of the house
were broken. Mr. Rothesay now began to look
troubled at the grave state of things, and accordingly
the police were set to watch as soon as possible. But
though on the alert for several nights, they failed to
discover who the offender might be. One evening,
without mentioning his intentions to the rest of the
household, Hubert determined to remain up that night,
in hopes that he might be able to fathom the mystery.
He retired as usual when the others did, but as soon as
all was quiet, he softly opened his door and went down-

stairs to the rooms below. There was no sound there. He then went round to the hall door, and gazed out into the darkness, but soon he was obliged to draw back, for the moonbeams began to quiver on the dark foliage of the trees, and he did not wish to be seen if any person should be lurking about. Presently there was a slight rustle in the bushes near at hand. At first Hubert thought it might be a stray rabbit from the warrens near, in which the fields abounded; but no! there it was again, louder than before, and his heart began to beat quickly, he was not in the least afraid, only excited, and very anxious to know what would happen next, for he had no doubt now that there was some one awake and stirring at that late hour besides himself.

Soon from out the black shadows of the trees, slowly and cautiously, emerged the figure of a man. Hubert knew instantly that it was none other than Winter. There was the slouching gait he knew so well, and in a moment, with a stride, he sprang to the path and confronted him. Winter fell back and tried to escape, but Hubert was too quick for him.

" Wait one moment," he said in a low voice—"I more than suspected this. And so it is to you we are indebted for those letters and broken windows, is it ? "

"Now, look you here," said the man—"if you peach on me you will rue the day to your life's end."

" It is of no use your riding the high horse in that style," replied Hubert; " you are now in my hands, and I have only to go to the police at once and give you into custody, which would serve you well right; but you shall have another chance. I know very well when a man is imprisoned his character is lost, but if you

will promise not to show yourself about the farm
any more, I will let the affair go no farther, so you
can please yourself."

The man shuffled with his feet, and hesitated.

"Come," said Hubert, "will you promise?"

"Well," replied the man, "if I must I must,
suppose," and, with a muttered oath, he walked slowly
away.

Hubert turned into the house, barring and bolting the
doors behind him. If he had followed his own inclina-
tion, it would have been to place the man in custody
directly, but he knew that just then it would not do any
good, but perhaps be conducive of an infinite deal of
harm, as there were others who had sided with Winter,
and if he had been imprisoned he would probably have
been regarded by the rest as a martyr or something like
it, and so make bad worse for Fern Glen. For some
time Winter appeared to have kept his promise, and
nothing was seen of him ; but, for all that, things did
not go on any better at the farm.

One bright morning, by sunrise, Hubert was abroad,
and might have been seen, if any had been awake at
that early hour, pacing with energetic strides over the
fields, now and then casting a look of troubled thought
on the ripening grain, as the question haunted him,
again and again, should he make peace with Giles
Winter and try him once more? He knew the man
with whom he had to deal ; he felt sure too that he was
doing all he could amongst the others to sully his and
his father's name, and was setting dark tales afloat
about the neighbourhood, which were readily taken up
and published abroad by the more ignorant inhabitants
of the village. Hubert was a proud fellow; he hated

the idea of condescending to such a man as Winter, and
ask him, as it were, to come back. Intelligent and
thoughtful beyond his years, he had taken upon himself
much of the management of the farm, for when quite a
lad his father, having great confidence in his son, had
brought him forward on all business occasions. Thus
the men in his father's employ had been used to come
to him as a matter of course, save this one man, who
had caused so much ill-feeling in the hitherto peaceful
village.

In vain Hubert worked early and late, lending his
aid with diligent, untiring energy, but it was of little
avail in so large a place. Besides, for days the sky, that
had been cloudless, now betokened a change of weather ;
purple clouds hung in heavy wreaths about the sun, as
he neared his setting, and the horned moon rose
surrounded with a misty halo peering through the
damp air with tearful gaze. Everything had this year
gone against the land-owners ; storms in different parts
of the country had been working sad havoc and ruin,
and why should not they visit Fern Glen ? So reasoned
Hubert as he strode along. In the meantime the men
who had left the farm spent their time in idling at the
public-house, and as it was nearing the annual feast
things grew worse instead of better. To the poor wives
and children all this meant destitution, for at the feast
there was generally some drunken brawl or other in
which their husbands or sons were implicated. It is
most probable that if Hubert had been brought up in
a town, this evil propensity on the part of these poor
villagers would have troubled him but little ; but having
lived in a lonely place all his life, and with but slight
experience of the temptations of worldly society, he was

ever contemplating how he could improve and raise those beneath him, and with whom he came in daily contact.

As the day drew on Mrs. Rothesay, with her woman's tact, without saying anything to her husband, made up her mind to go round and visit some of the wives of the men, and see what she could do to bring them to a better state of feeling. So, after the domestic affairs were over, she set off, in her little pony carriage, to call on one or two of her principal tenants, amongst whom was Betsy Hill, as neat and trim a little person as one could wish to see. Mrs. Rothesay was a great favourite amongst the villagers, and many of those who were sick and ailing in the wintry weather had listened with eager ear to catch the first tone of her kind and cheerful voice. It did her good to see the look of genuine pleasure that kindled in the eyes of the invalid when she uncovered the basket she carried, for in these poor cottages she seldom went empty-handed. Directly the chaise drew up to the low palings in front of the cottage, Mrs. Hill, with a bright smile, at once ran out, exclaiming,—

"Do come in, ma'am, you be always welcome!"

"Oh! Mrs. Hill, I did not know that you were washing, or I would have deferred my visit until another day."

"Don't say a word about that, ma'am," said the little woman, surveying her plump bare arms complacently. "I have just been and put a few things in soak, and our eldest girl can do almost as well as me now, for she's very handy."

"I am glad of that, Mrs. Hill; it is a great blessing when our children grow up dutiful and industrious,—

there are so many that are nothing but a sorrow and anxiety."

" It is, indeed, ma'am, and I have much to be thankful for."

At this juncture a cloud passed over her pleasant face. Mrs. Rothesay readily guessed what caused it.

" I am sorry your husband has left the farm," said Mrs. Rothesay, gently; "and, as winter draws on, I fear it will go hardly with him, so many mouths to feed too."

Mrs. Hill was silent a moment, and her eyes filled.

" I am thorough downhearted about it, I am, ma'am. I have talked to him, but it don't seem of no use. I'm sure it's cost me thousands of tears, it has !" After a pause, lowering her voice to a confidential tone, she said, " I'll just tell you what it is, ma'am,—I believe it's all along of that Giles Winter; he goes about and bribes and threatens all those who would give in, and I know of my own self there be some who would only be too glad to come back."

After kindly enquiring about the children, Mrs. Rothesay mounted her carriage, and when she had made a few more visits, she turned and drove swiftly homewards. The airy shadows of the birds flitted ever and anon across her path, and though all nature seemed to smile her a welcome as she passed, a feeling of sadness indefinable lay with a weight at her heart. What was to be the end of all this, who could tell ?

CHAPTER XI.

A LONG CHAPTER AND WEARY VIGIL.

CHAPTER XI.

A LONG CHAPTER AND WEARY VIGIL.

WHEN Meeta's lessons were over, the rest of the time was her own for the afternoon, and away she would run to her numerous favourites. There was Bella, a full-grown sheep, but she was still called the pet lamb, and directly she saw Meeta she was ready to run races with her, and would follow closely at the child's heels when she went to feed her rabbits, and look eagerly with her gentle eyes to see if there was anything for her too. Then there were Kitty and Nanny, the two goats which belonged to Meeta's sisters, and they bleated after her, pulling at the cords with which they were tied. Neddy pricked up his long ears, and waited to be caught with his basket of beans, and Meeta would stand on tiptoe to put the halter over his head and lead him away to the stable. When he was saddled she set off down the village to post the letters. Though so young, she was the busiest little maiden, her father's companion round the fields, and his special little confidante on all the affairs of the live stock. She would go with him in the gig sometimes to get clover for the colt, and Neddy would toss his head and trot away as delighted as his young mistress.

Now and then Kitty and Nanny came into the kitchen, and one day, most unexpectedly, when Mrs. Rothesay

and Alice were sitting at their needlework, they heard
the clatter of some little hoofs upstairs, and when Alice
ran out to see what it was, there was Kitty trotting
about on the landing, and Nanny making the best of
her way after her. Little Meeta laughed merrily when
Alice went up and made them follow her round the
house and down the stairs. They looked so odd as
they went, taking a step at a time, and Meeta wondered
whether that was the way the goats would go down
the mountains. Every evening Alice took them to the
meadow for the night, but they were so affectionate
that they would remain at the gate, after she had
left them, bleating for a long time.

When Alice returned she often went to her own
room for an hour's study ; it was prettily furnished, and
showed the refinement of her taste ; it was always in
perfect order, for she was never in a hurry, and thus
found time to keep everything in its place. The three
windows at the far end commanded an extensive view
of the surrounding country. She is standing now by
the open casement, the rays of the setting sun rest on
her fair face and long, waving hair. Thoughtful and
silent always, to those who did not stop to observe she
might have appeared indifferent, but nothing escaped
Alice ; she knew the trouble her father and brother were
in about the men, how much there was to be done, and
comparatively none to do it. If Hubert had only known
there was one who felt and cared so for him in all his
perplexities, it would have given him fresh courage ;
for sympathy, even if it cannot arrest an evil, is very
precious. But Alice, gentle and affectionate as was
her nature, could have given to those dearest to her
little idea of her real worth. She was the very opposite

of those favourites of society who, by their sweet manners and charming phrases, express so much; but often such natures are on the surface, and when they are better known disappoint those who find at last how unreal was that which they had taken for sincerity and truth of character. Except for little Meeta, the house at this time would have been very dull; the piano was seldom opened, and the pleasant musical evenings and cheerful conversations were now more of the past.

Mr. and Mrs. Rothesay had always studiously avoided discussing troublesome business affairs before their children; especially was it the case with merry, happy little Meeta, but their anxious looks were not lost upon the rest. It was very trying to witness the men standing about with their hands in their pockets when there was so much need of their aid, and to see too the sullen expression of discontent on so many faces.

All this time Giles Winter was plotting darkly over some plan to vent his ire and wrath on his young master. He had made up his mind to get out of the country, and if possible to sail for America, and now, after much thought, he had formed the wicked determination to decoy little Meeta away with him. He knew she was the pet of the family, and he knew, too, the love of the elder brother for his little sister, and he was certain that he could wound him far more deeply by injuring one he loved than if he were to do any violence to Hubert himself. His wife, who was an unprincipled woman, sided with him, and together they planned their cruel design. By dishonest practices the woman had obtained small sums of money from time to time, which now amounted to enough to pay their passage out, but they

7

had kept their secret so well that none had even the slightest idea that in this secluded village such a plot could exist.

It was drawing towards seven o'clock one evening, and Nettie had been very busy in the kitchen preparing cakes and pastry for the following day, as it was the eve of Hubert's birthday. Though twenty-two years of age, he was still styled Master Hu by the old servant, who had lived with Mr. Rothesay ever since he was a tiny child of four years. She had also sent little Meeta off on her donkey, to gather moss and wild flowers, of which there was an abundance in one of the covers, to deck the table for the ensuing day. Hubert, in the meantime, was musing to himself on the lawn.

Mr. and Mrs. Rothesay had gone to the town on business, and would not be home until late; Nettie could hear the footsteps of her young master in the garden as he brushed the grass, and every now and then she heard the sound of his low whistle, as he strolled by. As twilight deepened Nettie glanced at the clock, and was surprised to find it had grown so late. She at once began to bustle about to prepare tea, and feeling anxious that Meeta was gone so long, she determined to send Hubert after her. She was obliged to call twice before he answered, and when at length he stepped into the kitchen, it was with a face so grave that the old servant felt alarmed for a moment.

"Why, Master Hu, what is the matter? Is it about that bad man you are still fretting?—don't mind him, honey; he'll take himself off one of these days, I'll be bound, and you'll be well rid of him."

She little thought, poor soul! in what way, and how soon, her words would be verified. Then Nettie told him that Meeta had gone down the road to the cover more than an hour ago, and had not yet returned, adding that she was afraid there would be a storm before night, the evening was so oppressive.

"I will be off at once, and look after her," said Hubert. "She should not go so far alone, for there are so many twists and turns she might lose herself."

As he went by the kennel in the yard, his dog, Juno, began to whine piteously, and strain at her chain. Hu was going on, as he did not wish to lose any time, but the poor dog looked after him so pleadingly, with her intelligent eyes, that he turned and set her at liberty. Nothing could exceed her delight as she bounded and capered about him, and soon both master and dog were making with all speed to the cover.

A short time before, there might have been seen a man with slouched hat, and a gun under his arm, following the donkey and his unconscious little mistress. He watched her to the gate of the cover, saw her tie her donkey to the post, and after waiting until she was safely in, went after her. All arrangements for the intended voyage being settled, it had only remained for Winter to get the child away, but hitherto he had been unsuccessful, though day by day, towards evening, he had lurked unseen and unsuspected about the farm, until he had vowed in his anger to break into the house and murder the inmates. But now was his time! and crushing flowers and mosses beneath his heavy tread, he tore aside the brambles and brushwood, and prepared to follow her track.

Meeta heard the rustling swish of brier and bramble as they parted asunder, and the crackling of the twigs below, and looking round, as she knelt by the ivied roots of a tree, she beheld with terror the dark malignant face of Winter glaring down on her. She had always avoided him, and had ever been fearful of his gloomy downcast aspect, and now, with a shriek, she sprang up and strove to escape. But it was too late! already his arm was around her, and quick as thought, he bound a thick handkerchief over her mouth, to prevent her from calling for help.

In the meantime, Hubert was striding rapidly across the field, Juno bounding before, when suddenly she stopped short, and with outstretched neck and ears laid back, she uttered a low howl. "What is it, dog?" said Hubert, trying to reassure her,—"Ju, what is it?" With another fierce howl, she started with a spring and rushed away. A wailing wind rose suddenly amongst the trees, but in a moment all was still again. Hubert set off to follow the dog with a wild fear at his heart, and he vaulted the high white gate as if he trod on air. In another minute he found himself face to face with Winter and Meeta in his arms. The poor child stretched out her hands towards Hubert as he came up, but was unable to speak. The crimson blood mounted to his cheek and brow, and his deep eyes, generally so calm and quiet, flashed fire Putting strong control on himself, he mastered his passion, knowing that he stood alone and unarmed.

"Winter," he said hoarsely, "what business have you here? We have done nothing to provoke such treatment; if you must wreak your vengeance it shall be on me—but—spare my sister!" and, with a frame quivering with emotion, he was about to advance. In an instant Winter took up the gun and pointed it at him.

"Come a step farther," he hissed between his teeth, "and you are a dead man."

"If you don't put your gun down," replied Hubert quietly, "I will let the dog go."

All this time, he had his hand in Juno's collar, holding her back until her tongue lolled, and her eyes grew bloodshot, and as Hubert turned a moment towards the dog, endeavouring to calm her, the cowardly ruffian raised the gun to his shoulder and fired. He aimed only too well—Hu staggered back, his foot became entangled in the long grass, and he fell, striking his head against the gnarled roots of a tree, which stunned him for a time. His fingers relaxed their hold of the faithful dog, which now, with foaming mouth, sprang upon the brute's neck, and stuck her fangs into his throat. With a yell of pain; but still retaining his presence of mind, he flung off the dog and fired the contents of the other barrel to its heart. Raising the fainting child in his arms, he made for the road, sometimes stumbling and nearly falling in his eager haste, for a distant sound of wheels coming down the lane made him tremble.

He had been accused of theft, and now it might be murder that stood wan and pale at his door! The wind now rose in heavy gusts, sweeping by at intervals and moaning as it went, while between times the rain began to fall, and soon poured in torrents, dripping down with ceaseless patter from the laden foliage. But the damp and cold roused Hubert, as, sick and faint, he endeavoured to regain his footing and follow Meeta's track. Now and then he called for help as he staggered dizzily onwards. The wind and storm, as well as the rushing sound in his head, prevented his hearing the approaching sound of wheels. Still on they came—Mr. Rothesay, for it was he, had heard his cry, and had lashed his horse into a canter. Neither he nor his wife spoke; something fearful had happened, of that they felt certain, for the voice sounded hollow and indistinct as of some one in pain. Mr. Rothesay hastily took the lamp from the chaise, and with difficulty unfastened the wet lock of the gate, which slipped from his grasp. The flickering light threw weird shadows about his path, but on he went, guided by the sound of that almost inarticulate cry, which grew fainter and fainter, and at length ceased altogether; for, overcome by the drowsy giddiness that overpowered him, Hubert again sank down exhausted. He had striven in vain to wipe away the blood from his brow, which made his strained sight hazy and dim; but help was at hand, and presently the feeble rays of the lamp rested upon him.

Mr. Rothesay turned cold as he knelt down on the long damp grass by his side. "My boy!" he said, "what is this? what has happened?" With an almost super-human effort Hubert raised himself, saying in a husky whisper, "Go after Meeta!" Mr. Rothesay never knew how he managed to get back to the chaise; he told his wife as gently as he could that an accident had befallen Hu, and they must obtain help from the village. Without a word she took the reins and drove back. Alighting from the chaise she knocked at the first door she came to, and it was some few minutes before she could make the astonished inmates understand what had happened. But at last, under Mrs. Rothesay's instructions, a litter was prepared.

The ill news spread like wildfire; some remembered that they had heard the report of a gun, but thought it might be the keeper. All now began turning out of their homes, in spite of the weather, eager to give their assistance.

"Don't let them know yet up at the farm," said Mrs. Rothesay,—"it will come soon enough without that."

But the injunction came too late, for one of the farm lads, directly he heard what had happened, rushed off to the house with the news, and soon across the fields might be seen a white figure flitting along so fast that it seemed to be borne on the sobbing wind. It was Alice! They had become alarmed at the continued absence of Hubert and Meeta, and Alice was the first to see the boy, as she stood at the kitchen window eagerly looking out into the increasing darkness. Nettie ran to open the door, while Alice listened breathlessly. Hubert had been shot by some one—so much she gathered from the few sentences that sounded like a knell at her heart. Not waiting to hear more, she had rushed bare-headed into the rain, and scouring across the fields, in a few moments, guided by the light of the lamp, she stood at her father's side.

Kneeling on the sodden turf, she raised the head of her unconscious brother on her lap. "Alice," said Mr. Rothesay when he could speak, "how could you come? Go back, child, you will catch your death, you must not remain here." She raised her white face, pale to the lips, to his, and there was such wistful pathos in the mute appeal of her sad eyes, that he had not the heart to send her back.

Presently lights began to glimmer amongst the trees, and the men appeared bearing the litter. Gently they raised him, some of them those who had through Winter's evil example left the farm. They dared not look at their master: they feared his glance, and the expression of keen sorrow that they knew they must see on his benevolent countenance. Conscience-stricken, they did all they could, and they little knew until then how much Fern Glen Farm had been to them.

They had not realized before what they should feel to see one they had learned to respect and obey lying at their feet apparently lifeless. They now felt their conduct to have been shameful in the extreme; they could make no excuse for leaving Mr. Rothesay's employ, and wasting their days in idleness and drunkenness, as many of them had done. Their master's silence with regard to their behaviour, the pale form of Alice as she bent over her prostrate brother, unmanned some of them, ignorant and heartless as they appeared. Slowly the procession wended its way to the house, amidst a crowd of eager, curious questioners. Mr. Rothesay had looked forward to the time when, as he hopefully said, the men would see their folly, and before the cold of winter had set in peace would once more be established between them—but he little thought at what a cost !

At last they reached their destination. The door was opened by Mrs. Rothesay, how sadly altered in one short evening ! her face was haggard and pale, and there were dark hollows round her eyes. It was Alice now who came forward and gave directions ; it seemed as if the most quiet and gentle of the family had suddenly taken the place of those who had hitherto been the managers of the household. A new sphere had offered itself to Alice, and she entered upon it without question or comment. Noiselessly she hastened upstairs, giving orders where her brother was to be laid. Poor Nettie obeyed in a kind of heartbroken mechanical way, while Mr. and Mrs. Rothesay and Roger stood anxiously by. Alice raised his hair, now saturated with blood, and commenced sponging it away from his temples. The soft touch brought Hubert for a moment to a state of consciousness, and with a shudder he roused himself, and looked round with a wild, anxious gaze. Alice spoke to him in a clear low tone, " Hu, where is Meeta ?—try and tell us, dear, if you can." With a strong effort his lips framed Winter's name. But even that was too much, for memory and sense failed again. But it was enough for Alice—she guessed all now. Roger sprang down the stairway, and out of the house, running with full speed to the village to give the alarm. Soon he and the men with lanterns set off to Winter's cottage, but all there was dark and silent. They tried the door, but it was locked ; at length with some trouble it was forced open—there was no one there. It was plain enough the two were gone, but where was a mystery. Poor Roger hurried to the police station, to give a description of Winter and little Meeta. But the worst of it was, there was no telegraphic communication in this little country place, so that much valuable time was lost. Winter, who had been extolled by some of the villagers only the day before as "a plucky fellow, and one who wouldn't stand no nonsense," was now a " villain " and a " coward." Hubert had suddenly become a hero in their eyes; his unselfishness in thinking of his sister, when so severely wounded himself, was on everybody's lips.

In the meantime, Eva with trembling hands had saddled the horse, for there was no one else to do it, as all had gone off to see if they could hear anything of Winter, and she had reached the doctor's house covered with mud and foam. When the doctor arrived, he did all that could be done for his patient, but still Hubert

gave no outward sign that life was there. Mrs. Rothesay was so shaken that he ordered her to her room at
once, telling her that Hubert was in good hands, and
would have every attention. The table, spread for tea
more than two hours ago, remained as it was; no one
came into the room to close the shutters against the
storm outside; the Virginia creeper was torn from its
hold by the rough wind, and it kept up a ceaseless tattoo
against the window panes. In the still chamber above
the lamp flickered and glimmered, throwing deep
shadows where Hubert lay. The old eight-day clock
struck the hour below, which sounded sharp and eerily
through the silent house. But Alice scarcely noticed it
as she sat there, her small hands rigidly locked in one
another; and she looked indeed, clothed in white as she
was, like a guardian angel, keeping watch over one
intrusted to her care. She could well take the place of
a tender and skilful nurse, possessing the gift to which
even few women can lay claim; she had the loving heart,
and withal was so brave and steadfast, that if the call
had been, not by the bedside only, but to minister to the
sick and wounded on the field of battle, with the heralds
of death whizzing about her, she would have remained
faithful to the end.

The storm without began at length to subside, the
lattices ceased to rattle and shiver in the blast, for
the wind had grown weary, and wandered farther and
farther away, like a desolate soul seeking vainly for a
shelter. The pattering rain no longer flung its heavy
drops against the windows; but the soughing of the
distant trees kept up a monotonous sound, as of a far-
off sea wailing a mournful tune that knew no ending.

CHAPTER XII.

RONALD'S JOURNEY.

CHAPTER XII.

RONALD'S JOURNEY.

SOME few miles away, in a large well-lit dining-room, a youth was ensconced in a luxurious armchair, amusing himself with a little pug dog that lay snugly coiled up on his knee. His brown curls and sunburnt face were unchanged, except that his merry eyes might have become a shade or two more thoughtful. Ronald Haverleigh was perfectly acquainted with all the phases of the career which he had chosen for his own. Wet through, he had stood gazing out to sea when the night was black as pitch, ready to signal should impending harm be lurking near the vessel as she rocked and rolled, a mere atom, in the deep, driven before waves and wind. He knew, too, what storms at sea meant, when the shrouds were beating about him, and his hands were streaked with blood as he helped pull the slippery ropes and haul in. Or in a battle, still more dread, when shot and shell showered their fiery rain upon her deck, "of a piece with his ship," he had remained resolute and undaunted. It mattered not how sharp the ordeal he might be called upon to endure, the *Hero* herself must first become a mass of floating spars ere he would quit his post.

Still he loved his roving life so well that he would never have thrown it up for any less hazardous profession. On this particular evening, Squire Haverleigh and his son Bertrand had gone to the market, which was held in the small town before alluded to. Ronald's two sisters were in London, enjoying the novelty of town life. Jamie, who had always been delicate, had gone to the sea-side, in hopes that the salt breezes would make him strong again. Mrs. Haverleigh had retired to her room some hours before, and Ronald began to feel somewhat lonely as he awaited the tardy return of his father and brother. At length there was a knock and ring which roused him from his quiet musings, and presently the door opened and Bertrand entered, bringing a current of the damp air from without.

"Such a night!" he exclaimed,—"you may be glad that nothing calls you out. I left my father and came on, for I was tired of waiting. I do hope that he will not be long, for I am quite ready for dinner."

But when at length the Squire did make his appearance, there was no cheery exclamation as was his wont; on the contrary, he was grave and abstracted, and going to the window, for the shutters were not yet closed, looked out on the storm that was rising every instant.

"Father," said Ronald presently, "is anything the matter?" for the boy's quick eye detected at once the grave expression on the usually bright countenance. The Squire turned, saying somewhat abruptly,—

"Have you seen anything of your friend Hubert Rothesay lately?"

"Why, no, father; the last time I called he seemed almost too busy to notice me, but from what I could make out some of the men had behaved very badly, and they could not get fresh hands. I cannot imagine," he added, "why they kept that Winter on the farm so long."

" I heard a terrible piece of news," replied the Squire, "just as I was leaving the town to-night ; it seems that Winter, who is one of a thoroughly bad lot, has revenged himself by shooting poor Hubert, and carrying off little Meeta."

Ronald started from his chair.

" Impossible, father ! " he cried,—" I cannot believe it."

" It is true, I fear, for I had the news from a reliable source. I much wished to send one of my men down to ascertain the exact truth, but one would hardly like to turn a dog out while it rains like this."

" But Hubert is not dead, he cannot be dead ! " repeated Ronald.

"I devoutly hope it is not so," replied his father, " but that was the rumour."

" I can't sit here until I know the facts," returned Ronald.

" Where are you going ? " said the Squire as he moved away.

" To find out the truth—you will let me, father ? " he continued eagerly. " A puff of wind like this is nothing to me."

" Boy, you must be mad, it is the wildest night I ever remember. Come back, Ronald."

But he was off. Seizing his cap and a stout seaman's jacket from the stand in the hall, he opened the door and ran down the gravel walk towards the stables. He called the stable-boy, but he was nowhere to be seen. Quick as thought he got saddle and bridle from the saddle house, and put them on Snowberry, who looked round, it may have been with surprise, to see his master in such a hurry. Giving the pony a quick caress and a word of encouragement, he was in another moment slashing through the muddy roads in the teeth of the pelting rain and wind, at the animal's topmost speed. Generous and impulsive always, Ronald was one who generally went straight at a difficulty without awaiting second thoughts, and as he rode along his active brain was at work, considering how he could best help his friends in their trouble; and so he formed the determination of tracing the track of the missing child, and he saw no impediment which might not be overcome by promptitude and resolution. It was not long before he reached his destination, and opening the white gate, which stood out spectre-like in the darkness, he passed through. As he drew near the house, he could see shadows passing and repassing the closed blinds. " Oh ! then, he cannot be dead," thought Ronald, as with a lighter heart he quickened his pony's pace. As he went round to the back door, he saw a horse and trap standing in the yard, in the pouring rain, which he supposed belonged to the doctor. In a moment he had dismounted ; and gently raising the latch, he called in a low voice, "Is anybody there ? " Soon he heard steps approaching, and presently Nettie stood on the threshold, her eyes red with crying, and she could not for a moment command her voice to answer his hasty enquiries.

"Is he so very ill, Nettie ? " said Ronald sadly.

" Oh yes, sir, he doesn't know any one. To think only a few hours ago he was as well as you ! I little thought what a wretch that man was, or I should have trembled whenever my dear young master went anigh him—to think that we should have had any one such as him about us for so long; and dear little Miss Meeta, too, carried off no one knows where ! "

" Don't give way so," said Ronald, almost ready to cry

himself, for he was an affectionate, kind-hearted lad;
"we'll find her yet, and do say that Hu will get better!"

She shook her head sadly—"I can't tell," she replied;
"he has been ordered to be kept quiet, but the doctor
won't say one way or the other; he has not been able
to extract the shot yet."

"Can any one give me information as to where the
fellow went?" asked Ronald.

"London ways, so it is believed," said Nettie.

"Well," replied Ronald, "I'm off to seek Meeta;" and
before Nettie could recover her surprise or ask him
what he meant he was gone.

Once more mounting his reeking pony, he set off to
the station. The rain every moment was growing less
and less; the thick stormy clouds were gradually breaking, and lay in huge blocks along the horizon. There
was only one more train that night, and Ronald knew if
he lost it his ride would be for nothing. At length he
reached the station, and dismounting he called to one of
the lads who were lounging about.

"Here," he said, as the boy came up, "I want you
to ride this pony back to Haverleigh Park, and tell the
Squire—but stay a moment," he continued, "I had
better write a line, perhaps."

Giving the bridle into the boy's hand, and going into
the station-master's room, he hastily tore a leaf from his
pocket-book, and told his father of his intentions.
Giving the boy some money and the missive, he bade
him be sure and see that the Squire received it. He
watched him down the road, and then, taking his ticket,
Ronald paced the platform, restless and eager to be
gone. The porters were full of the sad story, and they
told him all that they knew, the universal impression

8

being that Winter had gone to London.   The hour struck
ten from some neighbouring clock, and the time he had
to wait seemed to him in his impatience interminable.
He could not see the signal, the night was so dark.   At
length there was a whistle in the distance, and soon the
roll of the coming train became more and more distinct.
Ronald thought that he never remembered to have seen
a train creep into the station at such a pace.   Once
seated he breathed more freely.   Divesting himself of
his soaking cap and coat, he prepared to make himself
comfortable ; but he could not sleep.   Visions of Hubert
ill, and perhaps dying ; of little Meeta, and what her fate
would be, drove all slumber away.   He tried to see
through the window, but could only catch the dim reflection of himself, and at intervals the black shadows of
the trees could be seen for a moment, and then they
vanished into darkness.

Soon he began to consider what his best course would
be when he arrived in London.   Once he had to change
trains ere he reached his journey's end, but all the
events of that night were so indistinct and unreal, they
appeared to him more as a dream than anything that
actually occurred.   At length he arrived at his destination ; the usually bustling station of London Bridge was
comparatively quiet at that hour.   Ronald scanned everybody who got out of the train, indeed all whom he saw,
with a curious gaze, but there was not the figure he
sought.   Seeing that it was fruitless to linger, he set
off to walk, as at present there was no cab to be seen.
The rain had ceased, though the clouds were still grey
and watery, but the waning crescent of the moon was
now visible.   Ronald made the best of his way towards
the Commercial Road, as there lived in that locality a

certain detective of the name of Adams, and to him he was determined to go for advice as to the best course to pursue. Being a sailor, Ronald had plenty of opportunities of striking up an acquaintance with one and another when he was in town, and the pleasant and jocular Adams, who was an intelligent and keen-sighted man, had long been one of the boy's friends. Threading through several narrow and intricate streets, he at length found the house that he sought. He went up hastily and knocked at the door, and, after some little time, he heard a firm heavy footstep approaching, and the door was opened by the very person he wished to see.

"Well met," said Ronald gladly, giving him a hearty shake of the hand; and he at once entered into his story, and why he had come there at that early hour.

CHAPTER XIII.

IN THE GREAT CITY.

CHAPTER XIII.

IN THE GREAT CITY.

BUT to return to Winter. Though wounded in the scuffle he managed to get clear of the cover without detection. He knew that when Hubert was found their first thought would be for him, so he hoped for the best. A miserable pony and cart were standing in the rain amidst the long grass, close to the hedge, where, in the increasing darkness, it could scarcely be seen. Taking a shawl from the cart, Winter entirely covered Meeta with it, tying it in such a way that she could not make any resistance. She tried to cry out, but was immediately silenced by Winter, who in a menacing attitude said, in a hissing voice, "Now look you here—if you make any noise, or give any trouble, I'll serve you as I did your brother—do you understand me? but if you are quiet you'll come to no harm." The terrified child was mute enough after this, and allowed herself to be placed in the cart without a word. Seizing the stump end of a whip, he belaboured the poor beast into a gallop, and away they shook and rocked over the uneven ground, for Winter feared to drive on the road, as the wheels might be heard. But the tempestuous night favoured his evil design, as every one who had a roof to cover them sought protection as fast as they could from the pouring rain, and there was not a creature to be seen. Winter pulled up some way from the station,

and springing out left the panting pony to shift for itself.
The woman, who had agreed to meet him there, was
ready waiting, and stepping forward silently showed
him the tickets, but no word passed between them.
She it was who now took charge of Meeta, and putting
her arm round the trembling child with a whispered
word of encouragement to "be a good girl," she lifted
her up and covered her with her shawl, telling her in a
low voice that if she was quiet, and did as she was bid,
she should soon go home again.

Winter pulled his hat low over his eyes, muttering an
imprecation on the train for keeping them waiting. He
was the first to hear the distant sound of its approach,
and when the porters came hurrying forward he slunk
back into the shade.

He watched his opportunity when the men were
putting some luggage into the van to get into a carriage,
shutting to the door and raising the window, while his
wife with little Meeta entered another compartment.

Hitherto they had escaped detection, though both felt
the greatest uneasiness until the signal had been given
and they had actually started. Soon afterwards, as some
of the villagers were hastening along the road in the hope
of finding a trace of the missing child, they discovered
the pony and cart by the wayside, which was recognized
as belonging to a market gardener in the village; the
poor starved pony was quietly grazing, and they agreed
to leave it and make the best of their way to the station.

Describing Winter, they asked the porters if they had
seen anything of him; one replied that the night was so
wet that he had not stopped to notice the people who
were waiting about, and another that he had seen a
man and woman, but he did not observe any child

with her. The two he saw did not appear to belong to each other, as they got into separate carriages.

The porter was sorry when he heard the sad story that he had not taken more notice, but he told the anxious group that as two people answering somewhat to their description had been seen at the station it was most probable they had escaped to London, as Nettie had said.

Dawn was approaching when the fugitives reached the outskirts of the Great City, and they immediately set forth in the direction of the East End, following circuitous ways to avoid any curious eye. At last, on reaching a dark archway, they were glad to stop and rest, and the woman taking some bread out of the basket she carried, they crouched down by the damp stone to eat it before they went farther.

Poor little Meeta by this time had somewhat recovered, though ill and weak with terror. Her senses were strained to the utmost, as she sat there silently by the woman, listening intently to their muttered conversation.

"'T'aint no use a-standing here," said Winter, at last ; " it won't be safe to get on her till night, for she don't sail till daybreak," and so they slunk away, the woman holding little Meeta by the hand.

Traversing several narrow streets, they at last stopped before a high, gaunt-looking house which stood in a narrow alley ; the rickety casements were many of them broken, and pieces of board or paper were stuck in, as the case might be. One or two ragged, miserable looking people were lurking about the door, that stood partly ajar. Winter pushed it open, and stumbled down the dark stairway. He explained to the man who kept the wretched place that he wanted a day's lodging, but should be off at night. The other eyed him somewhat

suspiciously at first, especially the woman and Meeta,
for the former wore a red handkerchief tied over her
head, and her black hair straggled from under it about
her sunburnt face, giving her a gipsy, vagabond appearance. But Winter put some money into the man's
hand, which seemed to satisfy him, as he allowed them
to pass into the room below. Winter then took a short
black pipe from his pocket, filled and lit it, and sitting
down on a broken-backed chair, began to think, if one
might judge by his countenance, for the scowl on his
brow deepened and grew more dark. And poor little
Meeta gazed at him with scared and tearful eyes, as she
crouched up in a corner of the room, for his face was
scratched and scarred by the previous night's encounter,
which made his appearance all the more fearful to her.

"Well," said the woman at last, looking round her,
"this *is* a place; sure you might ha' found somethin'
more decent; why, there ain't anither chair about the
place, nor table, nor nothin'."

"What would you have?" replied Winter, snappishly;
"you wouldn't have me go to a public, would yer, and
have the p'lice about my ears?" After this he again
relapsed into silence, and no more passed between them.
Presently he rose and went out, leaving the woman and
Meeta alone.

The former took off her shawl, and making a bed on
the floor persuaded the child to lie down. She was
kind to her in her way, but even she stood in awe of
Winter's savage and morose temper. Poor Meeta could
not sleep, though her companion, who had seated herself on the floor, was soon dozing, as Meeta guessed by
her heavy breathing. When Winter returned his face
betokened even more anxiety, and as he paced the

room he bent his head, as if listening to every sound
that could be heard without. He had accomplished his
errand, though not wholly satisfactorily, as will hereafter
be shown. Meeta watched him until she began to grow
drowsy ; but fearing to sleep in such company, she crept
towards the window, and tried to rub off some of the
dirt and cobwebs that hung there so thickly that she
could not see through it. All that she could make
out was that the room was below the ground, and everything appeared yellow and dim outside, for the atmosphere was still heavy with moisture, and a thick mist,
plentifully impregnated with smoke and soot, descended
like a cloud over the city ; it was one that could be felt,
for it hung cold and damp about the houses, and made
the room so chill that Meeta shivered and drew the
ragged shawl about her. She could hear the occasional
tread of stray pedestrians as they passed, but the footsteps were few and far between. There were not many
who ventured down there, or had any business in that
haunt of want and wretchedness.

At last she was startled by Winter telling her to
" let the window alone and come away." Hurriedly the
timid child went back into her corner, and soon from
fatigue she fell asleep. She still had on her little white
braided frock and velvet jacket, but her hat was taken
from her for fear of detection, and the shawl had been
wound round her to hide her dress. Her pretty sunny
hair was rough and tangled, and was thrust back into
the shawl, that was drawn partly over her head, and
those who had known her would scarcely have recognized
the once happy little Meeta in the pale sleeping child.

It grew darker yet ; still Winter paced up and down
the room, dim with the fog and wreaths of tobacco-smoke

that hung about it. He would fain, if he could, have got rid of the thoughts which troubled him. He fancied Hubert lying stiff and cold in the cover, and he could hear the dog's dying wail as she fell back into the long grass, and he pictured to himself how the scared birds would fly from the place, and how the moon would shine there, and show forth with her cold, clear light the victims of his wickedness. He wondered whether they had been found, and if the police were on his track. Supposing they should discover his retreat, and he should be hung for murder after all? Winter shuddered and wiped the damp drops from his brow. Oh! if he were only out of the country! and the room grew hot and close, as if it would stifle him, and yet he paced up and down—he could not rest. Hubert's still white face haunted him, and pursued him everywhere with a relentless fixed expression, which seemed to say that he should be brought to justice at last, in spite of all. He could scarcely wait for the long hours as they toiled wearily on; almost the whole day he had tramped up and down. At length he stumbled up the stairs again; it was now quite dusk; hazy lights flickered here and there in the damp air, though there was no light in the dark alley where he was. Returning, he beckoned the woman away from a rough group that had just come in, and the two withdrew, betaking themselves to the lodging-house keeper to obtain from him the disguises they required, so that they should not be recognized by any who might be on the look out for them.

When they came back Winter wore a rough coat buttoned to the throat; the collar was turned up so as almost to conceal his features, and in the place of the slouched hat he wore a broad-brimmed wideawake.

The woman's red handkerchief was changed for a rusty black bonnet and long cloak. The loud talking awakened poor little Meeta, and she looked with horror on the dirty appearance and inflamed excited faces of the people round her. At this moment Winter seized her by the hand; she uttered a faint cry, and tried to disengage herself from his grasp—she had not recognised him in his disguise.

"You didn't know me, eh?" said Winter, in a muffled voice : "all the better for us."

So saying he hurried her upstairs into the outer air. It was with a feeling almost of joy that the poor child found herself away once more from the heated and pestilent atmosphere of that wretched room. The fumes of stale tobacco and raw spirits made her sick and ill, and she was thankful to feel the air becoming every moment fresher as they walked swiftly along. What a maze of streets, and what a labyrinth of turns and twists, it seemed to the little country child. Soon, as they drew near the docks, she could distinguish by the lights the figures of sailors moving along, as they shouted and called one to another. Presently a sailor came up the quay; little Meeta could not see whether Winter had called to him or not, but it was plain he had been expecting them.

"This way," said the sailor, and soon the three with several others were scrambling up the ladder on to the ship. How dark it was—they could hardly see an inch before them! Poor little Meeta began to cry again, and her half-stifled sobs were at last heard by the sailor, who was leading the way. Turning round, he said in a rough but kindly tone, "What's the matter, little one?"

"Oh! don't mind her, she's only frightened," said the woman.

"Is she your little girl?" he asked.

"Yes," said the woman,—"she'll be all right when she's aboard."

The kind-hearted sailor was about to lift her in his arms, but the woman would not allow it, remarking that the child was well enough and able to walk.

Coming up, Winter told her in a low tone that "if she did not leave off crying he would put her overboard." A boy who was skulking off below heard Winter's words, and looked curiously after them. The excitement took off Meeta's attention, for what with the shouting and confusion, and the stumbling over ropes and tackling, she had enough to do to keep up, for it seemed every moment as if she must have fallen. Very soon after they had come on board, the woman and Meeta went below, while Winter waited to see that all was safe and that they had not been detected. Directly poor little Meeta had reached the cabin, the woman took a bun from her basket, pressing the child to eat, telling her that if she was a good girl and wouldn't cry no harm should befall her. But Meeta would not be comforted, but still sobbed in pleading accents to be taken home, and it must have been a hardened heart indeed that was not awakened to some feeling of pity by her distress. Even her rough companion began to feel some pangs of remorse, but it was too late to go back, and she could only promise the sorrowful child that she should go some day soon. Poor little Meeta turned away with a sickening sense from the food offered her, and presently, overcome by exhaustion, she fell into a deep slumber, the woman watching beside her. By daybreak the ship had weighed her anchor, and sailing out with the tide had soon passed from sight.

CHAPTER XIV.

THE SEARCH.

CHAPTER XIV.

THE SEARCH.

THE longest night must come to an end, and dawn was breaking once more over the hills surrounding Fern Glen. It seemed to Mr. Rothesay that the troubles which had come upon them were more than he could bear. Moodily he strode up and down the dining-room, which appeared all the more dreary in contrast to the flowers, that hung rich and blooming in the vases on the table, plainly indicating the loving preparations for the approaching happy day that was to have been, but now would never be. Then his thoughts went from Hubert o the missing child. Where was his little Meeta ? He scarcely dared to think, and unable to stay within any longer, he hastily took up his hat and quitted the house. As he neared the park gate he caught sight of Neddy waiting to be let in ; he had been entirely forgotten amidst the confusion and sorrow of the preceding night. As Mr. Rothesay approached, the poor beast laid down his ears and hung his head, as if he knew what had happened. Not very securely tied, he had broken away, and had endeavoured to find a better shelter from the rain under the trees. Mr. Rothesay led him back into the stable, taking off the wet saddle and bridle, giving him some corn and a fresh bed of straw.

By six o'clock the truant men appeared. As for their

9

master, he felt for the time that he could not bear to see them, their presence caused him to realize more vividly than ever the sorrow that had come upon him. But their conduct showed how penitent they were. He briefly told one of the men to bring the barrow and follow him, as he wished to find the dog. She was not in her kennel, and there was little doubt that she had been with Hubert on that fatal night. The two then walked slowly on, and passing through the gate of the cover, in a few moments they came upon poor Juno, lying stiff in the damp, sodden grass. At Mr. Rothesay's command the man put the dog in the barrow, a grave was dug in one corner of the home field, and a piece of wood placed there to mark the spot. Just as he returned to the house, Nettie came breathlessly downstairs to tell him that Master Hu had recovered consciousness and had spoken. Mr. Rothesay went at once to his son. Hubert was lying with his hand in that of Alice, as if he would not let her go out of his sight; he seemed to have but a dim recollection of what had happened to himself, though by his continually recurring to Meeta, he showed that he had not altogether forgotten the sad events that had occurred. Nettie tried to pacify him by telling him that he must not trouble about her, as Master Ronald had gone to bring her back. But Hubert could not recall to mind who Ronald was. How strange it seemed to that once happy and cheerful family, the darkened room and utter stillness. He was so weak, poor fellow! from loss of blood, that life appeared to hang on the merest thread. The doctor was very silent and grave, and all the answer that they could get to their anxious enquiries was that he had a fine constitution, and that he might rally if the internal hæmorrhage could be stayed. He was ordered as much ice as he could be got to take, and no trouble was spared to forward his recovery. Friends and neighbours made daily enquiries about Hubert and his little sister, and some were most anxious that Mrs. Rothesay should have a nurse for Hubert, declaring that they knew of one or two most efficient persons. But that was one thing that Alice would not hear of, and Mrs. Rothesay particularly shrank from the thought of having strangers about her. Thanks to Nettie's unvarying attentions, she was now better, and was able to comfort Alice and take part with her in her weary watching.

On this early morning, too, Ronald had set out on his search for Meeta. After he and his friend had break-fasted, away they started, Ronald declaring that he felt fit for anything. In a short time they found themselves in the precincts of the Victoria Docks. Detective Adams told him that he had no doubt that very soon the tale would be in the papers, and that Winter would not long be an unknown personage in a secluded village, but a public character; that his crime and the decoying away of the child would soon be made known all over London. He was suddenly startled by an exclamation from Ronald, who started off at full speed. The detective quickened his pace, but could not see him anywhere. After a short time, however, Ronald appeared again. " I be-lieve I saw him ! " he panted, his flushed face getting more excited as he spoke—" a man with a slouched hat, and now I have lost him." Adams at once became deeply interested, and his firm-set features and sharp penetrating eyes looked like one who was not going to be trifled with. They hastened their steps, searching every-where,—down by the black grimy wharves, under archways, and wherever an unlucky personage could possibly hide away, there went Adams and his indefatigable companion.

At length they were obliged to give it up. It would not do to ask any questions, at present everything depended upon secrecy. Ronald was dreadfully disappointed, and lurked about the Docks as if he were infatuated; and at last, from sheer fatigue, he was obliged to give in, and seek a night's repose. Adams, meanwhile, had communicated with some more of his calling, and he told Ronald that he must keep up heart and hope for better things on the morrow. Late that night Adams turned out again and took his way to the Docks; he knew there was a ship about sailing for America, and he thought there might be a chance that the man had taken his passage in that vessel. He watched the sailors, who were busily lading, and asked for the captain, but no one seemed able just then to give information as to his whereabouts. He also inquired when the ship sailed, but could obtain no satisfactory answer; he then told one of the crew that he was looking out for some one, who might for aught he knew have taken his passage on the ship. With some difficulty the detective stumbled up the ladder, and looked keenly at a knot of people who had assembled on deck; he scanned every lounger, and no one escaped his penetrating gaze. He went hither and thither, down the narrow passage leading to the saloon, peering into cabins whenever he could get an opportunity, but none did he see answering to the description of the man he sought. He stayed for a moment watching the huge bales as they were swung one after another by the creaking crane into the hold below, amidst the shouts and hoarse cries of the crew.

At length, with a last look around, the detective made for the ladder again. It was now so dark that he was obliged to feel his way, and he had just reached the ground when Winter and the woman, with poor little Meeta, amongst several others, were about to ascend. Quick as thought, Winter's escort pushed them back into the shadow, while the detective halted for a moment, taking a sharp survey of the group which stood waiting to mount.

"Now then," said a voice, "make way there," and Adams found himself roughly pushed aside; but he sturdily stood his ground as far as he could, watching as the people hustled and jostled each other up the steep ladder, endeavouring to grasp the hands that were stretched out from above to help them on deck. But Adams saw none answering to Ronald's description, and Winter was still favoured under cover of the thick murky night. When he was safe on board, he little thought what a narrow escape he had had, nor did Adams dream how near he had been to the man he sought.

Baffled still, he sauntered farther down, looking from between the labyrinth of masts, lit up here and there with faint gleams of light, that flickered in the dusky night. The mist that had hung about all day became heavier still, and Adams felt that he was as far off the scent as ever.

Ronald was not wrong when on the early morning of the day he fancied he had seen Winter, who on leaving the lodging-house had taken his way down to the Docks, to ascertain exactly where the ship lay, and also to see one of the crew whom he knew, and who had promised to stand by him and see him safely on board; so far it was satisfactory. On his way back the sharp fellow was on the alert, ready to fly should he be unlucky enough to find anyone on his track; he had caught sight of Ronald before he saw him, and was off like a shot; he knew of winding streets and intricate turnings of which Ronald had no idea, and the race was from the first a most ill-matched one. Winter had lived a vagabond roving life before he settled down in the quiet little village of Fern Glen, and was quite familiar with the loathsome haunts and back slums of the great city. But he was ever silent and sullen, and the innocent villagers knew nought of his past history.

Early next morning they were again on their search. "We will keep about the Docks," said Adams,—"the fellow will very likely get clear off from the country if he can."

Ronald followed silently; his spirits began to get low, and the thought that it was of no use, and that he had come on a fool's errand, pressed itself heavily upon him. He could not go back without the child, he had so set his heart upon finding her. Adams continued talking to him; he had it pretty well to himself now, and he explained very minutely the for and against of their quest. But it sounded such a mass of complication that Ronald scarcely listened, and soon was wholly taken up with his own sad thoughts.

As they came up once more to their old haunt, they discovered that the ship had sailed that was loading on the previous night. Thus another day was passed in fruitless search, and Ronald spent his evening in writing home.

All this time there was little change in the big house at Fern Glen. When Roger came from school, he wandered about like a restless spirit; he could not bear to go into the room where his brother lay—he was so altered that it seemed to Roger that he could not be his strong brother Hu at all, and the sad events were to him as a strange bad dream, from which one day or other he would perhaps awaken. The house too was so hushed and still; there was no merry child's voice or ringing laugh, no tripping small footsteps up and down the staircase. He would not go near the labourers, or help, as Hubert used to do—he felt in a rage and passion with them all, and was ever telling his father how he would have acted, declaring that if he had been in Hu's place he "would have set the dog right on the man, and wrenched the gun from him." But Mr. Rothesay told him he was sure Hu had done rightly, and that if he had gone to work as Roger said, he must have been killed, for he had no chance, not having any weapon with him. But Roger would not listen to reason, his heart was sore with anger and sorrow, and there was none to speak to or comfort him. Eva had too much to do in the house; then there was the poultry, and poor Meeta's rabbits, all needing the same attention as if nothing had happened, besides the numberless duties belonging to a country farm. There were the goats bleating in the yard, looking in vain for their gentle mistress Alice; they came into the kitchen, following Eva everywhere she went, until they got so in her way, that she was obliged to drive them out, and fastening a collar and cord to the neck of each, she took them down into the meadow and shut them in. How well she remembered but a short time ago they had so enjoyed running them into the field, and what fun they had had, harnessing them to Meeta's pretty little goat-chaise. The contrast was so great compared to the sad present that the tears came into Eva's eyes as she walked slowly back; but she resolutely drove them away, it was no time for fretting, there was too much to be done.

The afternoon sunshine lay in long lines of light over the undulating fields, and nature with her thousand charms was adorning all around with brightness and beauty; but she had no power to drive away the pale shadow of sickness, that darkened the house with its depressing influence. Poor little Meeta was not only missed in her home, but the labourers in the fields, too, felt strangely lost without her, for they always knew what time of day it was when they espied her coming down the lane on her donkey, to post the letters at the other end of the village; nor was it only they, for Mr. Cecil often found himself on a Sunday looking towards the pew in search of the earnest little face he knew so well. The high square pew in the grey old church had been very frequently empty since that sad time; the rays of the summer sunlight streamed in at the stained casement above it, but there was no sunny hair to reflect back its brightness. The cottagers in their clean gowns, and little children with their white socks and well brushed shoes, hastened to church as the cracked bells jangled out that time was getting on, and the sight of the children reminded poor Mrs. Rothesay so strongly of her own little daughter that she felt quite unable to go out or see any one.

The big house-dog Ben knew as well as possible that there was something wrong, and would lie all day in the yard with his nose buried between his paws. He evidently missed his little playfellow, who used to pat and hug his huge head so lovingly. Mrs. Rothesay had a painting of Meeta, taken when she was quite a tiny child, with her arms round the dog's neck, and a very pretty picture they made. Ben considered himself the child's protector, and little Meeta had implicit trust in his faithful care—indeed Mrs. Rothesay never used to feel anxious about the child when Ben was by; but at the time of which we are writing, when Meeta was lost, he was getting very old, and seldom went beyond the yard. When any friends came to enquire after Hubert, which was of daily occurrence, the old dog, who was generally on the watch, would immediately get up, and wagging his tail to show his pleasure, would take a part of their dress in his mouth, leading them towards the house as nicely as any child could do. But if strangers came whom he did not know, he would bark loudly, so that Nettie might come quickly to ascertain who was there. So you see he was very wise as well as useful.

Towards evening Mrs. Rothesay prevailed upon Alice to leave the invalid with her, and go out for a short time. Very reluctantly she left the room, and slowly went downstairs. She had not crossed the threshold of the door since that dreadful night that now seemed so long ago. The freshness of the soft air fanned her aching head, and soothed her at once. She took her way along the fields to the pool where her favourite willow grew, that had been planted and reared since she could remember, for it did not seem so long ago when it was but a small slip, and now its cool green reflection vied with its companions of larger growth in preventing the water from being dried up by the heat of the sun, so that there was always plenty for the hot thirsty cattle, when they came for a refuge from the troublesome

insects or to ruminate beneath the shade. Alice could
hear the rippling gurgle of the brook near at hand; that
at least was the same, and the perfect calm by which
she was surrounded endowed her with fresh strength
and courage to endure. The sweet forget-me-not looked
up unchanged from amidst the long rushes and grass,
and around the stem of the bending willow the purple-
eyed vetch wound its tendrils; wild water-mint, silver
weed, and convolvoli flourished beneath the hedges, and
from amongst the sombre green the bright leaves of
the elder and wild rose peeped here and there. August
was fast giving place to the rich September, with her
warm and glowing tints; the sun was sinking now behind the blue hills in the distance, and the waning daylight warned Alice that it was getting late. So quickly
culling a few of the pale blue stars from the brook, she
turned homewards. Mrs. Rothesay met her at the door.

"I am so glad you have come, dear," she said, "for
Hu is so restless—I am sure he misses you."

In another moment Alice was again at her post. Her
evening walk had so refreshed her that she felt more
hopeful than she had done since Hubert's illness. "I
have only to trust," she said to herself, "and all will be
well." Yet poor Hu still lay in the same semi-conscious
state that he had been all through. Alice seated herself
by the bed-side, and placing her arm beneath the pillow,
she gently altered his position, and drawing his head
towards her she brushed the damp threads of hair from
his brow. As she sat in the silence, her thoughts
wandered away to her little sister, and presently she
began in a low voice a song that she and Meeta used to
sing together. Hu turned as if listening, and lulled by
the soft sweet sound he fell asleep.

How long Alice had remained thus she scarcely knew, when the door softly opened and the doctor entered. He came up to the bedside, and looked at Hubert intently for a few moments. Glancing at Alice he said, "Your brother is much better; he must not be disturbed on any account, everything now depends on perfect quiet." Alice's heart gave a great bound ; how she longed to run and tell them all downstairs ! but she would not move. The doctor placed a small table beside her with all she might require, and then withdrew. Hour after hour passed; her position grew cramped and irksome in the extreme, until at last she seemed to have lost all sense of feeling. The clock below struck twelve, yet still he slept, and Alice almost trembled, and longed for him to wake. At length a stray moon-beam fell clear and pale across the floor, and as it gradually glided from sight, a star or two glimmered and twinkled in at the window, and Alice felt comforted as she gazed up at them, so quiet and ethereal they were ; storms could not ruffle their majestic serenity ; clouds might hide their light, but they shone still with their eternal radiance unchanged, unchangeable.

Suddenly with the first streak of grey dawn Hu awoke. He turned and addressed Alice by name. "Have I been ill ? " he asked. " Oh, I forgot," he murmured, as recollection returned, " I remember it all now."

" You have been very ill, darling," said Alice, " but you must not talk, or you will make yourself worse. It cost her a great effort to answer him thus quietly. Weak with watching, it seemed now that that for which she had prayed had come to pass she was unable to bear it. It might have been that their troubles had come upon them so suddenly that she had borne them hitherto with a kind of dumb submissive grief too deep for tears.

"One word, Alice," said Hu, entreatingly,—"tell me if you have had news of our dear little Meeta?"

"We are expecting to hear from Ronald every day, dear," said Alice, evasively. With a deep sigh he turned away.

"You must not fret, Hu," said Alice, "for what with you and Meeta, it has made poor mamma quite ill. Do all you can to get well and strong," she added, more cheerfully, "and before long I have no doubt we shall hear of Meeta."

"Ronald is a good fellow," said Hubert; "he is always ready to help any one in trouble, I feel his kindness all the more as I cannot go myself."

"If the doctor hears you talking like this," said his sister, "he will give me a bad character, and you will have to put up with another nurse."

"But I won't!" replied Hubert, and his decided, almost irritable, tone persuaded Alice more than anything else could have done that he really had taken the first step towards recovery. Thus a ray of sunlight had at last broken through the dark clouds of sorrow that had swept over Fern Glen.

CHAPTER XV.

NEWS.

CHAPTER XV.

NEWS.

NEARLY a fortnight passed away, and very gradually Hubert continued to improve. He still
had to remain in bed, at which he fumed so that he was
an infinitely more trying subject to deal with than in
the first stage of his illness. But Alice did not heed
his complaints—they were indications of returning health,
and that was all she cared for.

Up to this time no trace of the missing child had been
discovered, and Ronald Haverleigh wandered about dis-
consolately with no aim, having nothing to do. He
generally betook himself down to the Docks, wishing
devoutly that something might arise that would unravel
the mystery. He was all the more dispirited as he had
received an intimation that the *Hero* would be sailing
shortly, and he feared he would have to give up the
search, and it might be that poor little Meeta would
never be found at all. Slowly and sadly he took his way
along the quay, and not noticing particularly where he
was going, he stumbled and nearly fell over the prostrate
figure of a boy, who lay coiled up fast asleep.

"Halloa, youngster!" said Ronald cheerily, "don't be
frightened, I did not mean to disturb you," as the boy
sprang to his feet with a scared expression on his
pinched countenance. As Ronald looked at him, he became
interested in the boy at once, for in spite of his ragged

and dirty appearance he perceived he was dressed in the garb of a sailor.

"What is your name?" asked Ronald, "and how do you come here?"

"My name's Jack Linton," said the boy, "and I wants

work—I ain't got no home nor a farthing to get a bit of bread with."

"Poor fellow!" said Ronald compassionately, "and yet you have been to sea, I should imagine—where is your ship?"

"She sailed these two weeks back," said the boy.

"I see," said Ronald smiling, "you have backed out."

"I couldn't stand it no longer," he replied, rather sullenly; "they wer' so bad to me all the time I was aboard; the crew was always ordering me about, and I got more kicks than ha'pence,—no more sea for me!" said the boy emphatically.

"Why, you foolish fellow!" exclaimed Ronald, "you did not think that all you had to do was to go staring about with your hands in your pockets; if you had stuck to your work and your ship you would have been fit for something by this time."

"It's of no use now," said the lad, "she's gone."

"What was her name?" asked Ronald.

"The *Sea Swallow*, and she was bound for New York. She'd some emigrants a-board too, for as I was a-managing to scramble out without the other fellows seeing me and telling the cap'en, I saw a desperate-looking cove coming down the gangway, and I thanked my stars I wasn't going to sail along with him far."

Ronald started with a thrill. "Supposing—what was he like?" said Ronald, trying to speak calmly, "was he dark and scowling?"

"Do you know him?" said the boy, looking up hastily.

"Answer my question," said the other, seizing his arm.

"You won't blab on me," said the boy, beginning to

10

whimper; for he had seen the gold lace on Ronald's cap, and he thought at once that he might have something to do with the ship.

"Indeed I will not," said Ronald earnestly, "I will get you work and stand your friend if you will tell me all you know."

"Well, he wer' a wery dark man," said the boy, "and he wore a long rough coat, which covered up his face a good bit, and he had on a broad-brimmed hat."

But the difference of dress did not baffle Ronald, for he felt persuaded that the man the boy had seen was none other than Winter.

"Was he alone?" asked Ronald.

"No, there was a child and a woman along with him, sir."

"A girl?" asked Ronald quickly.

"I think it were a little gal, and—oh! yes, I remember now, she was a-crying; she was all muffled up in a shawl or somethin', and he told her that he'd throw her overboard if she didn't shut up," said the boy, "and she wer' wery quiet arter that."

"What time was it, should you think, when you saw them?"

"Between ten and eleven," said Jack, "but I ain't sure, it wer' wery dark."

"Look here," said Ronald, "here is a shilling for you to get some food; I will have a talk with you afterwards about yourself, but I must not stay now. That man you saw was a rogue and a scoundrel, we have been hunting after him this fortnight past," and slapping the boy on the back he exclaimed in a hearty voice, "You are a brick, old fellow, and you will never repent what you have told me!"

Then, with a parting injunction to be there when he came back, he was off at full speed, leaving the astonished boy staring open-mouthed after him. Away flew Ronald, banging up against pedestrians who were hurrying another way, some of whom turned and scolded him for his carelessness, but he was too excited to heed. Arriving at his destination, he burst open the door, and shouted out the news to Adams, who was just coming down-stairs. On entering the room he told Ronald they would hasten over their breakfast while they discussed the best course to pursue. But Ronald was too eager to eat much ; he told Adams that Winter had got clear off to America, and what was more he had managed it on the very ship that he had been searching. The detective listened very quietly until he came to that, and then burst into a torrent of invectives. " He was not going to be made a fool of by such a fellow as that, he who was famed for his sagacity and foresight, who had unravelled mysteries, who had brought, as it were, the light of his bull's eye to bear upon the most cunningly contrived schemes, but he would be even with the fellow yet, or he was much mistaken ! "

By the time they had finished breakfast, Adams was thoroughly roused, and had worked himself into a white heat, to Ronald's infinite satisfaction. They walked rapidly down to the Docks, to enquire after any steamer likely to be going to America, and if they could be fortunate enough to fall in with one, there would be no more uncertainty, as the bird was "as good as bagged," as Adams expressed it. They would arrive there first, of course, and would arrest the man before he had.time to leave the ship. They found there would not be a steamer going to America for two days, which was most disappointing, but it was no good to lounge about idle, as the time would not pass away any the more quickly for the wishing; so the first thing Ronald did was to keep his promise to the poor sailor-boy to whom he owed so much. He had no difficulty in finding him, for he was wandering about the quay, evidently on the look out for his benefactor; and while Adams was gone to make the necessary arrangements for their expedition Ronald informed the lad of the sad story of little Meeta. He then told him that he intended sending him to his father's house, where he could take the place of errand-boy, at least for the present, and the poor lad gladly accepted the offer.

His story was only one amongst a thousand such. It seemed that the boy had a stepfather who had ill used him; his mother was dead, and all that he had been able to earn by selling matches or running errands was got from him by his father, and at last in desperation the poor boy fled from the only place he had ever known as home, and sought to better his condition in London.

He had tramped all the way from Liverpool in search of work, but owing to his ragged appearance and having no character, people would not trust him, and so he wandered about homeless and hungry, until he was taken on board the *Sea Swallow*, and Ronald knew the rest.

Ronald now bade poor Jack follow him and get " rigged up a little." Being a sailor, and used to manage for himself, he had plenty of experience about things of which other youths in his position would have been perfectly ignorant. He at once took him to a warehouse that he knew, and soon no one would have recognized in Jack the ragged boy who was sleeping hungry and cold at the Docks a short time befo re. From the warehouse they took their way to the station. Ronald gave him his father's name and address, and placed him under the care of the guard, who promised to look after him. The poor boy thanked his benefactor again and again for his kindness, and could scarcely believe in his good fortune. He promised to do his best if the Squire would only keep him. Bidding the lad farewell, with a pleasant word of encouragement, Ronald was off to his lodgings to acquaint his father with what he had done. He told him how much he owed to the boy, as through him he was enabled to trace poor little Meeta's whereabouts, and he begged him to keep the lad, at any rate until his return, after which he began a letter to his friends at Fern Glen.

It had been the hardest part of all for Mr. Rothesay, that he had not been able himself to seek his little daughter, but he could not possibly go from home while Hubert remained so ill, nor could he leave his men at the busiest time of the year, and Roger was too young and inexperienced.

Ronald was the very one of all others in whom Mr. Rothesay would have the most trust, and he thought if any one could succeed in the search it was he. Ronald explained why he had not written before; he did not wish unnecessarily to raise their hopes until he had some good news to communicate, which had come at last. He spoke of his voyage to America and his quick return; of little Meeta, and how he should never be satisfied until he had restored her to her home. It would have done him good could he have seen the joy his letter produced when it arrived the following morning. The news at first was kept from Hubert, as they feared the excitement would throw him back.

After some little discussion it was agreed that Mr. Rothesay should go to London, his wife declaring that he could well be spared for a night, at any rate. He was longing to see Ronald and know from himself all he had to tell, for boys are not exactly good scribes in the general way; they do not enter into minute particulars, and there was much still the family wished to know.

Mr. Rothesay set off at once by the midday train, and on his arrival took his way to the Commercial Road. As he neared the house Ronald, who was standing at the window, caught sight of him, and came out at once to give him a welcome.

"I half expected this," he said smiling—"I thought you would come if you could only get away." He then enquired anxiously after Hu.

Mr. Rothesay told him that he had been very dangerously ill, but now they had every hope that with care he would get well in time. With regard to the child, he still felt very low spirited about her; knowing how timid and nervous she was, he feared to think what she must have suffered; but Ronald spoke so hopefully, and seemed so cheerful about her, that he, too, began to take heart and feel encouraged. As they went together to the hotel, Mr. Rothesay enquired at what time he would start in the morning.

"By seven o'clock sharp, I hope," replied Ronald. "I will be there nothing preventing," he answered, and so with a warm shake of the hand they parted for the night.

That night, like many others that had preceded it, was a feverish one for Mr. Rothesay, for in his slumbers he was ever pursuing his missing child, and as the weary time dragged by the night seemed to him never-ending.

Every now and then he started awake, hoping to find it day; but only darkness reigned, and amidst the never-ceasing roll of distant wheels he heard a neighbouring clock strike one! while trembling on the air he caught the far-off sound of the big bell booming forth the hour with solemn tone from its airy height in the great cathedral.

When at length he again fell into an uneasy sleep his dreary search began again, and he was threading his way through interminable streets, or rocking over the broad sea in search of his little Meeta. Suddenly in his dream she would appear before him, gliding rapidly away some distance in advance, and when after infinite exertion he came up with the child, like an *ignis fatuus* she vanished from his eager hands. Thus he woke and slept and dreamed again, but ever with the same result. Mr. Rothesay was thankful when the bright rays of the morning illumined his chamber, and put to flight the shadowy visions of that restless, miserable night; and by six o'clock he gladly quitted the hotel, taking his way to the Docks, and pacing up and down waited for Ronald, who at length appeared, coming briskly towards him, saying with the bright smile and cheerful manner peculiar to him, " Why, you are early, you are before us all this morning."

The tide being out, the passengers were all hurrying down to reach the tug which was awaiting them. What a holloaing there was as the sailors shouted one to another! Mr. Rothesay was almost bewildered as he looked on at the noisy, busy scene; but Ronald was used to it, it was nothing to him. Presently Adams came up, accompanied by two other officers, provided with a warrant for the man's apprehension—all were disguised in plain clothes. Mr. Rothesay could not find words to express his thankfulness to the kind-hearted boy who had done so much for them in their time of need and distress. Placing his hand on his shoulder, he said in a voice full of emotion, "When you return, my boy, you must come and stay at our house; my wife and I will welcome you as a son, and you will ever have a place in our gratitude and affection. We can never repay you, Ronald." Mr. Rothesay was not the man to make long speeches, but what he did say had the more effect, for it was heartfelt and sincere.

Ronald was delighted to receive his praise, and seizing his hand he replied, "You must not say a word about gratitude; what you have just said has repaid me more than enough for all that I can do."

At length the signal was given, the black smoke began to pour from the funnel, the heavy wheel to beat a tattoo in the water, the foam and spray dashed and flew in the wind, and the good vessel was at last fairly off, ploughing her way in real earnest. It seemed as if she, too, knew of little Meeta, and was in haste to be gone to bring her back to her dear father and mother, who had been watching for her so long and so sadly. Mr. Rothesay stood shading his eyes with his hand to catch a last glimpse of her ere she passed from sight, and Ronald waved his cap from the deck until the land and the watching figure could no longer be seen.

CHAPTER XVI.

THE VOYAGE.

CHAPTER XVI.

THE VOYAGE.

WHEN the villagers heard that news had come at last of poor little Meeta, there was universal rejoicing. Squire Haverleigh's domain was made the rendezvous of all the gossips in the neighbourhood. Jack's story was considered as good as a book, and he had repeated it so often that he began to enlarge a little, and dilate on the fierce words and looks of Winter until his tale was fast partaking of the horrible. Supposing that the wicked man had been as good as his word, and had thrown poor dear little Miss Meeta overboard; but that was too terrible to be thought of, so day by day they impatiently waited and watched for the sequel.

Gradually the good news had become known to Hubert; his quick eye perceived in a moment the difference, the settled shadow on his mother's face had suddenly lifted, he had not seen her so cheerful for a long time, and he looked inquiringly at his sisters as they came in and out of his room. Mrs. Rothesay was busily arranging some flowers on the table, while Hu reclined on a couch by the window, gazing dreamily at the clouds as they floated past. Some were grey and gloomy as they almost imperceptibly arose from the dim horizon, but when the sun's rays caught them their sombre tints were transformed, and they sailed along in the ether like

fleecy islets of snow, while a sky of liquid blue showed
forth here and there from the infinite beyond them.
But Hu was not thinking of the clouds just then, he
was longing to ask if there had been any news of his
little sister, but feared to do so lest it might bring the
shade back again to his mother's face, as it had always
done whenever he mentioned her name; so he made up
his mind to wait until Alice was alone with him, she
was so quiet nothing ever seemed to ruffle her. He
was ever as he lay there recurring to the events of
that terrible night, picturing to his morbid fancy how
the cover would look when the sun arose on the following morning, and how his rays would glint in and out
amongst the dripping leaves of the trees, and rest on the
dark ivy beneath, stained by something else besides rain.
He dwelt on the fate of poor Juno : he did not ask if the
dog was saved, he knew too well. Some might think
his silence on the subject betokened forgetfulness, but
it was not so; he mourned truly for the dog that had
been so faithful to him, and tortured himself with fancy-
ing that he might have managed better, until his
brain throbbed with feverish excitement. If he had
only taken his gun, he thought he might have been able
to save Meeta, and perhaps the poor dog also, but these
harrowing regrets were worse than useless, they only
retarded his recovery. Presently Alice came in with
her work in her hand; softly she glided up and leant
over the back of his couch.

"Hu," she said, "what will you give me for a little
piece of news that I have to communicate ? "

He turned quickly, exclaiming, " Oh, then, it is true !
I fancied it must be so from the appearance of you all
this morning, but I dared not ask before."

"You are not mistaken, dear Hu," said Alice, fervently. "We trust now, if all goes well, that our darling little sister will again be restored to us." And then Alice read to him Ronald's letter.

Though it somewhat unnerved and excited him, it was infinitely better for Hu than to be continually dwelling on the sad past.

"We can never repay him," he said, when his sister had finished. "If he had been our own brother he could not have done more."

"It is just like him," replied Alice. "He is such a generous fellow," she added, kindling. "There is nothing that he finds a hardship so that he can alleviate the troubles of others, his heart is so thoroughly in all that he does." The emotion betrayed in her voice was so unlike her usually calm manner that Hu looked at her enquiringly, but Alice was scanning with far-off gaze the distant horizon, where earth and sky blended in one.

And all this while the good steamer was ploughing her way through the water; the ripples began to rise into waves as every moment bore them farther and farther out to sea. Ronald felt the first breath of the salt breeze as he stood on deck, and his fine eyes sparkled with exultation as he felt himself wafted out on the surging deep he loved so well, which began to roll and rise and sprinkle the deck with spray. He bared his brow to the fresh breeze as it blew past; he watched the sea-gulls as they flew hither and thither, eddying round the vessel, then skimming along the waves, their wings glancing like silver in the sunshine. Many of the passengers noticed Ronald with interest, his bright handsome face, ruddy and burnt with exposure to sun and wind, his energetic movements and manly bearing, until suddenly catching their

admiring gaze with a quick flush he would turn abruptly
away.   Perhaps it may almost excite surprise that he
should have taken so much trouble about a little child
that was no relation to him; but setting aside the friendship that had existed between the two families for so long,
he was very fond of children, especially the merry, happy
little Meeta, in whom he had discovered thoughts and
feelings beyond her childish years.   She had been out
with him many a time nutting in the woods; he had
often let her ride his pony, and she had been his little
companion on many a skating expedition.

But besides his love for the child, Ronald possessed a
thoroughly unselfish disposition as beautiful as it is rare.
He was the favourite son of his mother, and beloved by
all who knew him for his honest and kindly nature ; he
had a good word for the humblest, and never passed any
by proudly or indifferently as unworthy of notice ; and
yet none would have thought of addressing him with
familiarity, for with all his freedom and ease of manner
there was a certain dignity which commanded respect.

The captain and he were very good friends, and Ronald
soon made known to him the reason why he was on
board, and who were his companions.   As the vessel
neared mid-ocean, thick storm-clouds began to pile one
above another, hanging their wreaths of white vapour
like a veil over the black blocks of cloud beyond, their
edges tinged with a copper hue, and there was every
appearance of rough weather.   The sea tossed and
moaned as if conscious of a coming battle with a gale,
sea birds began to multiply and shine white against
the murky clouds, and every day the atmosphere grew
denser.   Soon there was a sudden lull in the elements,
lightning ever and anon trembled and quivered in the

far west, and many of the fearful and more timid of the passengers went below. Daylight was fast waning, and suddenly gave place to darkness—there was no twilight for that night; the lightning now became more vivid, it shot in zigzag flashes of blue light across the heavens; as if charging earth and ocean with fixed bayonets. The clouds opened and closed again with the noise of a hundred cannon—the vibration seemed to make the brave vessel shudder as it died away, as if she knew what was before her ere the land should come in sight and her work be done.

To Ronald it seemed that she must know her responsibility, and that all on board depended on her for their safety. The furious wind shrieked and screamed through the rigging, and he was obliged to hold on by the rail to prevent himself from being blown away. He watched the murky mass of clouds as they flew in streaks before the wind, while the huge waves drenched the deck with showers of spray; the electric fluid became continually more lurid, glancing above and around in flashes of scarlet fire, the chains in the vessel rattled and shook as the light played about them; the rain began to fall in torrents, dashing against the ship in its fury, and the elements all conspired together in one deafening roar. As to the steamer, she began to strain and creak, and at last she made no way at all, as the waves mast high buffeted her on every side.

The captain managed to get close to Ronald, who was sheltering himself behind the boards and tackling. " We've sprung a leak ! " he shouted, " and if we don't mind we shall have the fires extinguished, and the boiler may burst,—we must not alarm the passengers."

One or two scared faces appeared at this moment, but

the captain ordered them down again. Without waiting to hear more, away went Ronald to join the men at the pumps, and after a few hours, to their intense relief and thankfulness, the storm began to lessen in fury. The wild scream of the sea-birds became more indistinct, and at the first approach of dawn there was a partial lull in the tempest.

Ronald came up again wiping his hot face, and the captain advised him to go to his berth and lie down, which he was glad to do after his exertions, and rolling himself in the blanket he was soon asleep.

All through the next day, though rain had ceased to fall, the wind still blew furiously, and there was a heavy chopping sea, but early on the following morning, the welcome rays of the sun came struggling through the clouds of mist and vapour, as he rose from the east above the boiling crested waves which still foamed and swelled from the recent gale. The clouds fell away and turned from grey into rose and orange, and the gloomy surface of the water sparkled and glistened once more in the ethereal tints of the morning skies. The brave steamer began to take heart again, and was soon on her way as if nothing had happened. The damage she had sustained was not so great as was at first feared, and the ship's carpenter made all as tight as he could for the time, until she should put in for more thorough repair. In the calm which followed, however, fog wreaths began to rise, and the air soon became so thick that the vessel was enveloped as in a sea of cloud, and ever and anon the fog horn blew. Song and music were suspended during those hours of semi-darkness. In silence and anxiety the passengers looked forward to their deliverance. Every eye was weary with straining its gaze through the dense damp veil which hung pall-like over all. At length, to their relief and delight, the steamer once more emerged into the light of day, and songs and gaiety and laughter were again in the ascendant, and as she went rapidly along, the sea was as a huge and shining mirror, which shivered into myriads of dazzling fragments as she cut her way through. Now and then a solitary ship could be seen on the horizon sailing to other lands, and as evening approached the clear atmosphere betokened a rich and gorgeous sunset, as he descended towards the west like a ball of fire.

Ronald stood silently on deck looking intently at the exquisite scene. What a contrast to the preceding days! The sky, the ship, the water, all partook of the roseate hue; below the sun, radiant clouds of dazzling light floated in the deep violet-blue ether, and it seemed to Ronald that he could discern another sea between the clouds, stretching far away into the infinite. Its ripples were of molten gold shaded with pale green, and ships of silver floated on the golden waves; their sails, like the white wings of a sea-bird, gleamed in the glorious beams of the red sun, and it was as if the gates of Paradise had opened for a moment, shining with the jewelled lustre of opals. Rapidly the sun sank beneath the long glistening billows, and the rich colours gradually faded, until there was nothing but a streak of violet cloud lying in a pale blue sea of sky in the far-off west. Though unnoticed before, hidden by the glow of the sunset, now that all was grey the evening star arose, bright, steadfast, beautiful, gazing with clear eye from heaven to earth. Its gentle influence tranquillized Ronald as if his mother's cool soft hand had been laid on his head, and a whisper like an angel's voice breathed

around him. Boy-like, he had scarcely ever thought of
heaven or death,—they were too far off, he had not
realized their meaning; the present life to him was all,
and that other life in comparison appeared so pale, so
cold, so colourless, that he turned from the thought of
it, and was glad to feel that his present existence, so
buoyant and so full of life, was his still; that nature
with her rare and exquisite colouring, that foreign climes
with their spicy productions, the sea with its sands
and its shells, were for him to enjoy for many, many
years to come. But now as he stood there, his earnest
eyes fixed on that evening star, it seemed as if his life,
with all its hopes and fears, was going out in one long
yearning glance to meet that other life of which he knew
so little. Just for a moment the beautiful earth, the
grand and awful sea, were as nothing compared with
that other land, of the golden shores of which he had
seemed for a brief space to have had a glimpse. Presently with a sigh he turned away with an expression
on his thoughtful face unknown to it before. Ronald
could never be again exactly as he had been. In that
solemn moment the hidden meaning of existence, its
greatness and responsibilities, was dawning upon him,
and he was becoming conscious that the present was
but the prelude of that which was to come, infinite,
eternal!

He was suddenly roused from his reverie by the loud
tones of the captain, proclaiming that land was in sight.
Ronald was by his side in a moment, and as he looked,
there, sure enough, he saw a low dark line in the horizon.
The far-off expression on his countenance had vanished,
he was all life and animation, talking and explaining
to this one and that, as they came up scanning the

distance with their glasses. Presently he was joined by detective Adams and his companions; they told him that they must be on the alert, for it would be, as Adams expressed it, "rather a close shave," for the weather had sadly hindered them, and the ship might be in now for aught that they knew, and then it would make "a sight o' trouble."

But every moment brought them nearer, and Ronald's eyes were fixed on that long line which began to get more and more distinct, and at length the lights in the distance could be discerned as they glimmered and twinkled from amongst the dark foliage of the trees.

Anchoring at Staten Island, the passengers awaited the arrival of the health officer; but it was too late for him, so the vessel remained there all night, to Ronald's disappointment, and restless and impatient he paced up and down the deck, eager for the coming day. By sunrise, however, they set sail again, and the good steamer was making the best of her way to the wharf of the Anchor Line.

CHAPTER XVII.

ANTICIPATION.

CHAPTER XVII.

ANTICIPATION.

SUNDAY had once more come round with its solemn calm, when the singing of birds seems to sound more hushed, and from the neighbouring churches the soft chime of distant bells sighs and swells on every breeze. On this day Hubert was all excitement, for he was to leave his sick-room for the first time and come downstairs. The green blinds in the drawing-room threw a cool shade on all around, making the flowers and ferns on the low sill all the brighter from contrast. The rich petunia looked in at the window, filling the room with its sweet fragrance, and clinging with its tendrils to the creeper and ivy that hung down like a curtain, while quivers of sunlight threw the reflections of the leaves in dancing circles on the floor.

Soon Hubert was reclining in an easy chair in the pleasant room, a shadow of what he had been; but though he was still weak he was free from pain, and the life so nearly lost was slowly but surely returning. With what eagerness did he now look forward to his recovery, when he would be able to go out of doors again, and enjoy the pure air that blew over the pine-clad hills! He had never known before the value of health; he took it as a matter of course that he should be possessed of strength and muscular power above many.

But as he sat there powerless to move a step by himself, never had life been half so dear to him as now.
He had sometimes complained of his isolated circumstances, of his not being able to travel and see more of
the world, but now it seemed to him that, if he could
walk once more, as he used to do, over the fields and
amongst the familiar scenes he loved so well, he would
be completely satisfied.

On this same Sunday morning, too, Ronald Haverleigh was standing on the quay, accompanied by Adams
and his two companions. All were looking anxiously
out to sea. The *Sea Swallow* had been expected in
the evening before, but the ship at present had not
made her appearance. Hour after hour passed, and
still they stood and watched. Some few idlers, who
fancied by their manner that something unusual was
afloat, loafed about and watched too.

It was early in the afternoon when, at the farthest
point, the sun's beams rested on what appeared at first
like a tiny cloud, but it grew more and more distinct,
until there could no longer be any doubt that the sun
was shining on the sails of a ship. The four on the
quay watched her breathlessly, and as she came nearer,
her black hulk was plainly visible. "I could swear
that's she," said Adams in a whisper. Nearer she came,
and sailors were rushing down with the well-known cry,
" Ship ahoy ! "

The two detectives, with Adams in their wake, got
the handcuffs ready to pounce upon their miserable
prey the moment they caught a sight of them. The
three men were presently walking rapidly up the gangway, and stood where everyone must pass that left the
vessel. Ronald fell behind, and was obliged to hold by

the rail to steady himself; his heart beat with anticipation. Could it be that at last he should find little Meeta ? He could scarcely contain his joy as he thought of it. The people pushed and crowded past him, and the boy's bright, sharp eyes scanned and measured each one.

"Look, look !" he exclaimed in a whisper to Adams, "there he is !"

Notwithstanding the disguise, Ronald knew him in a moment, and there, sure enough, was the man they sought. His face was well-nigh hidden by his hat, and he walked with a hunted air, as if he knew an avenging spirit was dodging every step. A woman followed after with a little girl. They had just reached the opening, and Winter was preparing to set foot on the ladder, when Adams sprang between him and it, saying in a stern, clear voice, loud enough for all to hear,—

"Stop ! I must have a word with you first, my friend. You are my prisoner. I arrest you on the charge of shooting Mr. Hubert Rothesay, with the intent to do grievous bodily harm, and also with the decoying away from her home Miss Rothesay, of Fern Glen Farm, ——shire."

The man fell back ; his knees knocked together, and he seemed too utterly dumbfoundered to offer any resistance. The woman shrieked and endeavoured to extricate herself as another of the detectives laid hold on her arm, and said quietly, "You are our prisoner also."

Ronald hesitated a moment. Could that poor child be Meeta ? Her cheeks used to be fresh and rosy, and this child was pale and thin ; it seemed impossible. Mastering his agitation he called her, " Meeta, don't you know me ? "

She had been so wholly taken up with the horror-stricken countenances of Winter and the woman, and in trying to comprehend what Adams said, that she had not noticed any one else. Hearing her name she turned quickly, and with a cry of joy rushed into Ronald's arms.

The crowd that had so eagerly been pressing out of the ship now endeavoured to return. Adams was beset with questions on every side, and there was much ado before anything like quiet could be restored. Ronald beat a retreat as quickly as he could from the noise and confusion. Meeta's arms were strained about his neck as if she would never let go.

"Take me home, Roy," she murmured, "take me home!" and then the strained grasp relaxed, and she fainted.

Ronald walked rapidly down the streets with the unconscious child to the hotel where he was staying. Mrs. Walton, the landlady, was looking out for him, having heard his story. Directly he reached the house she did all she could to comfort him, explaining that it was only a fainting fit, and that the little girl would be better by-and-bye.

There were many anxious faces amongst the waiters and maids at the hotel, who beset the door of the room when Meeta was carried in, and everybody wished to do something for her. Very unwillingly Ronald relinquished his precious burden to Mrs. Walton, and she told him that when once the poor little girl was placed in bed, and she had given her a restorative, he should come and sit with her, and Ronald was obliged to be satisfied.

Kind Mrs. Walton herself carried poor Meeta upstairs while Ronald betook himself to the telegraph office, to convey the good news to those at home. When at last Meeta opened her eyes she was lying on a small iron bedstead, and some one was bathing her forehead with vinegar. As she became better she was placed in a refreshing bath of warm water, which imparted to the poor little child a delicious sense of rest and comfort, though her mind was still clouded and bewildered, and she endeavoured in a dreamy way to understand where she was, and how she came there. When at length she was laid in bed her eyes wandered round the large, pleasant room, to the white drapery shrouding the bed and the window, and everything looked bright and clean to the sick, weary child. Suddenly recollecting herself she turned timidly, and seeing a stranger, was beginning to cry.

" You must not fret, love," said Mrs. Walton in a kind voice, " the young gentleman will soon come back to you."

Meeta looked up shyly as she spoke, and it was such a good-natured, homely face, that her fear was set at rest at once. Little by little she was encouraged to tell her tale, and all that had befallen her since she was taken from her home.

The kind woman's eyes filled as she said, " Poor child ! you are very young to have suffered so much, but," she added more cheerfully, " you will soon see your papa and mamma now, I hope."

Just at this moment a quick step was heard mounting the staircase, and Ronald entered. Meeta held out her arms towards him. " Don't leave me, Roy," she said, " don't go away ! "

" I won't leave you, dear," he answered, taking her thin hands in his own ; " I have come all the way to take you back."

" And poor Hu ? " she whispered, looking anxiously at Ronald.

He assured her that when he heard last the news

was so good that he had no doubt Hu would soon be all right again. "And now," he added, as he stationed himself by the bedside, "you must not talk, little Meeta; I will be spokesman, and you must listen."

The child answered him with a bright smile, and she was very glad to lie and be silent. The blissful rest after the days and nights of suffering she had undergone was delightful in the extreme, and she lay with her eyes fixed upon Ronald, as if she feared to let him out of her sight, and when at last the weary, heavy lids closed, he was still before her as she slept.

When she awoke an hour after, he was sitting there just the same in the waning light. Starting up in bed, she exclaimed, "Oh, Roy, have you remained here with me all this long time? How tired you must be!"

Her face was flushed with sleep, and she looked more like the little Meeta of old times. Ronald laughed, as he told her it would take a great deal more than that to tire him.

"You shall have some tea now," he said, "for I have just found out that I am desperately hungry, and we will have it together up here; would you not like that?"

Meeta was delighted. In answer to the summons a pleasant-looking girl came into the room, and quickly wheeled a table to the bedside, setting thereon a tray with cups and saucers. Having made the tea, she presently came in, bringing some chicken and omelette from Mrs. Walton's own table.

Ronald insisted on feeding Meeta, which so amused her, that her happy, ringing laugh was heard once again, which did her more good than anything else. The fresh butter and new milk seemed now the greatest luxuries to the poor little half-starved child, and every now and then she would stop to embrace the noble-hearted boy who had done so much for her; Ronald was determined that no sad recollections should cast a shade over the first evening of their meeting, and he was himself so elated with the success of his expedition, so full of fun and lightheartedness, that it was the merriest tea-table imaginable. When the tea-things were cleared away, Meeta begged Ronald to tell her all he could of the loved ones at home, and she wondered when they would hear of her safety. He told her that the good news would soon reach them, as he had telegraphed, and they had only to wait one more day, when they would be ploughing their way back to old England.

Then Meeta gave him a description of the miserable day in the lodging-house, and of the unhappy time she spent on board the *Sea Swallow;* and how she used to sit and watch the waves breaking into spray, as they rolled up the sides of the ship. She told how the sea-birds flew about her until she fancied that perhaps they knew and pitied her in her sorrow and loneliness, and how angry Winter was if she was noticed or spoken to by any of the passengers, as was sometimes the case.

" The wretch !" exclaimed Ronald hotly; " well, he will soon be in gaol, and then he will be the right man in the right place, I reckon."

" Will he go to gaol ? " said Meeta, opening her eyes.

" Of course he must," replied Ronald, " when he tried to take poor Hu's life, and has treated you so shamefully. If such men as he are not brought to justice, folks would be in danger of their lives every day. The man who can oppress and ill-treat any one weaker than himself is a coward. I always used to fight the fellows at school if they bullied the little ones."

Mrs. Walton reappeared and told Meeta that she had talked long enough and must go to sleep, so Ronald kissed her good-night, promising to be there the first thing in the morning. The next day, as little Meeta was still weak, it was thought wisest for her to remain in bed, but the thought of the coming journey prevented her from troubling over that, though it was irksome to a naturally active child to lie for so many hours. Towards evening, however, at her earnest request, and as she appeared much better, she was allowed to sit up, and Mrs. Walton, wrapping her in a warm dressing-gown, placed the child in an easy chair by the window, where she could see the people passing and repassing in the street below. Her heart was full of joy at the thought of the coming morrow, and she could think and talk of nothing else.

There was not much sleep for Meeta that night, and the next morning she was so excited and eager to be dressed, that she was up long before the good landlady thought it prudent. However, she let the child have her way, as she knew she would never rest until she was ready to start. Ronald had posted off to buy Meeta a hat, and Mrs. Walton was quite amused at his going on such an errand ; he, however, told her to suspend her judgment until his return. Some time afterwards Ronald appeared with the prettiest little sailor hat, tied with blue ribbons, and "Princess" inscribed in gold letters, which Mrs. Walton declared did credit to his taste. Little Meeta was very pleased, and Ronald told her that as she had proved herself so good a sailor she was worthy of a hat of the proper cut, and he had determined she should be made all "tight and taut" for her homeward journey. When her long, light hair had been brushed, and she had on her clean white dress and new hat, she was as pretty a little girl as one could wish to see, and Ronald was pleased enough to take charge of her. As Meeta was too excited to eat, her kind hostess brought her a small bag well stocked with sandwiches and buns, telling her that she would be quite ready for them by-and-bye, after which she kissed her farewell, and little Meeta looked up with such a sweet smile, and embraced her so affectionately, that Mrs. Walton quite loved the child, and begged Ronald to let her know when they arrived in England. Ronald promised, at the same time heartily thanking Mrs. Walton for her kind attentions to little Meeta when she stood so sorely in need of a friend. Many of the travellers staying at the hotel had heard Meeta's story, and numerous were the cheering words and good wishes which followed her on her departure; for none could help looking with interest on the child with her heavy, wistful, blue eyes, and pale little face; but there was now a happy expression, for all the sorrow had passed away, and it effects time would soon efface. At length, with her small hand tightly clasped in that of Ronald, she was making her way down towards the quay, and soon they caught a glimpse of the vessel as she rode at anchor in the harbour. It was a lovely day, and the sun shed his beams of welcome on the sunny head of the joyous child as she went along. Towards evening Meeta was standing on deck with her kind friend. "Oh, Roy!" she exclaimed, clinging to his arm, "what should I have done if you had not come to take me back?"

"Ah! little Meeta," he said tenderly, "you cannot think how I sought for you, until I began to fear it was of no use."

"Don't you think, Roy," said the child, after a pause, "it must be something like heaven ? Why, it seems to me that if I were there first I should be always looking out for those I love, and when I saw them like I did you when you came on the ship, I should just spring into their arms, and lead them in." He turned his bright face for a moment towards her, and the serious far-off expression came back into his eyes as he said in a low voice, " I don't know much about these things, but in a sense it may be so. Very soon I shall be obliged to part with you all, and be off again, and in yonder heaven there will not be any need to say farewell ; but, whatever happens, the chief thing is to be prepared. You would think that sailors of all fellows would feel this, they have such awful hairbreadth escapes, but they don't, and often there are the very worst specimens of humanity amongst a ship's crew."

Then Ronald pointed out to Meeta the misty haze across the water, and how if she watched she would presently see the moon rise above the clouds ; and soon her disc began to appear, and looked so large and red that Meeta thought for a moment it must be the sun, and that he had made a mistake and risen before his time. The stars shone clear and bright overhead, and Meeta was never tired of looking at them. How different now it seemed ; she ·saw everything from another point of view. In her voyage, so desolate and alone, to an unknown land, she had scanned the sky and stars through bitter tears, and all around had partaken of the sadness that was weighing so heavily on her young heart ; but now all was changed, and each day brought her nearer and nearer to her own dear home and loving parents.

CHAPTER XVIII.

REJOICINGS.

CHAPTER XVIII.

REJOICINGS.

EARLY one morning there was a good deal of interest and excitement amongst the villagers of Fern Glen. Neighbours from the various cottages stood in groups conversing together and looking eagerly down the road, for on this particular day little Meeta was expected home. Just in front of the white park gate leading to the house was constructed a triumphal arch, and there were men still on the ladder busily putting the finishing touches, while Eva and Alice stood below with some damask roses and ivy. And this was not all. Squire Haverleigh intended giving a grand children's party in honour of Meeta's return, which was to take place the following evening, and the day after was to be a tea-drinking and a dance in the barn for all the villagers, which meant a regular holiday. For when the Squire did give a treat he entered into it thoroughly, for he was a large-hearted man, and there was no lack of hospitality. There would be long tables filled with buttered rolls, plum-cake, and tea and coffee in abundance.

Eva and Alice, with Mildred Haverleigh and her sister, were busy from morning to night wreathing the bare walls of the barn with evergreens, until it looked inside more like a bower than a barn.

Early in the afternoon, Mr. and Mrs. Rothesay came to the door ready to start for the station. Eva had placed a favour on the horse's head, so that he should appear in holiday trim as well as the rest. He pricked up his ears as if he knew there was something unusual, and was in haste to be gone. Soon they were whirling away to the little station.

When they reached their destination, Mr. Rothesay walked up and down the platform, his countenance expressing eager anxiety as he looked and listened for the coming train, and Roger, leaving the horse in charge of a boy, darted down to the gate upon the first signal of its approach. On came the train, and those few minutes seemed longer to poor Mrs. Rothesay than any in her life before. As it stopped at the station, the door of one of the carriages was flung open, and out sprang Ronald, while little Meeta held out her hands, but before she had time to speak she was caught in the arms of her father, who carried her at once to her mother.

It was almost too much for Mrs. Rothesay. "My little child!" was all she could say, and she strained Meeta to her as if she would never let her out of her sight again. Then both parents turned to Ronald, and he was amply repaid as he received their fervent expressions of deep thankfulness for all his kindness.

Squire Haverleigh, who was there also, could not blame his son, though he had gone off in such a "harum-scarum fashion," as he termed it. He could not help admiring him when he saw what a hero he had become, and how everybody pressed up to shake hands with him, for there was quite a crowd at the station. Mr. Cecil was amongst the number, and warmly expressed his approbation.

"Well done, Ronald!" he exclaimed, placing his hand

on his shoulder, "you have acted nobly, and nothing done out of pure sympathy for our fellow-creatures shall ever go unblessed."

That was a proud moment for Ronald, and most thoroughly did he deserve the well-merited praise. Mrs. Rothesay begged Squire Haverleigh that he would spare his son to return with them that evening. "He was perfectly willing," he said, "as far as he was concerned, but as Ronald would so soon be off to sea again, he must not stay long, for his mother was anxiously looking out for him, and she wished him to spend the rest of his time with her."

The Squire told Mrs. Rothesay that the brougham would call on the following day for herself and Hubert; he would not hear of a refusal, Hu must be one of the party; there would be a couch for him to lie upon, and if the noise was too much, he could easily be moved to another room.

"Mind you go to bed early to-night, little Meeta," continued the Squire, "so as to be ready for the party tomorrow."

Ronald and Roger then mounted behind, and they drove away. As the chaise neared the village it was followed by numbers of the cottagers, who waved their hats and handkerchiefs, and shouted a welcome. Children with smiling faces stood at the park gate, holding it open for them to pass through. Nettie was at the door ready to receive them, and as they drew up, the old servant hurried to the chaise. In a moment Meeta was in her arms, and was carried into the hall, followed closely by her two sisters, and she scarcely knew which to embrace first. And now once more the unbroken family party sat round the well-spread board. Little Meeta nearly broke down when she first saw Hubert looking so thin and pale, and the tears stood in the eyes of both brother and sister as they kissed each other; but Ronald, with his merry ways, determined there should be nothing of that sort on that night of all others, and by his ready wit and clever tact turned the channel of their thoughts from all that had a depressing influence, and soon Hubert became as cheerful as the rest. Yet if any had occasion to be dispirited it was Ronald, for this was probably almost his last evening on land for many a long day to come, but, brave boy as he was, he uttered no word that should in any way sadden them.

Little Meeta sat on her father's knee looking the picture of joy and contentment. Both parents were quite surprised to find the child so well. They had been prepared to see a little, pale, delicate face, instead of which there was a bright, sunburnt, rosy little lassie, who had been the favourite of all on board, who had tripped after the sailors, and endeared herself to them by the interest she took, and by her artless remarks on all that she saw. All were much gratified to hear of Mrs. Walton's kindness and care of the child, and Mrs. Rothesay took the first opportunity of writing to thank her for her hospitality, and every year the good landlady received a substantial proof of their gratitude.

As evening approached, Mrs. Rothesay told Ronald that she must not keep him any longer, as they would be looking out for him at home. Little Meeta clung to him as if she would never let him go, and Hubert seized his hands in both of his, while Ronald burst into a merry laugh as he tried to disengage himself from them ;

but Roger, who was going part of the way with him, was in a hurry, and inclined to be cross, as he declared that "they would not be there before nightfall at that rate."

When they reached the hall door they espied Jamie on his pony, leading his brother's by the bridle, coming quickly up the field.

"Roy," shouted James, "mother thinks it is time you were home, she has been watching for you this long time. So with a hasty farewell, hoping to meet the next day, Ronald mounted his pony, while Alice and Roger walked beside them to the gate.

"Why," said Jamie, "you have a splendid arch; I should like to have helped, but I have not been well, and they would not let me out of doors, but I begged off to-night; I do hate to be shut in all day."

"Poor Jamie!" said Alice, laying her hand on his arm as he rode along, "then you must not linger, less you should make your cough worse." The advice was needed, for the boy's sweet face looked transparently pale to-night.

"Give my love to Meeta," said James, as he set off at a canter down the lane.

That night Meeta was ensconced in her own little white bed, and her mother sat by her till she fell asleep. Early on the following morning she awoke to hear the first sleepy twitter of the sparrows beneath the eaves; she could hear the lowing of the cattle in the distance and the cooing of the pigeons in the yard, and she could scarcely wait until it was time to get up, so eager was she to see her pets from which she had been parted so long.

When the child was dressed, she ran down into the dining-room, where Nettie was laying the cloth for breakfast.

How the kind woman folded her in her arms with tears, for she had never forgiven herself for allowing Meeta to go alone to the cover.

It seemed to the child as if she had been away for months instead of weeks; her little heart was full of thankfulness, and never had her home been so dear to her as now. Nettie had missed her terribly, for she was generally her little companion in the long twilight evenings, when she would perch herself upon the sill of the open window, framed in by the ivy, and, with her kitten in her arms, she would sing hymns in her pretty child's voice to Nettie as she worked.

Soon the child was running down the fields. How delightful was the dewy grass, and how sweet the breeze which raised her floating hair! How blue the hills looked in the clear atmosphere, while nearer at hand she saw the sage-green foliage of the trees in the cover, and with a sudden shiver she turned away.

After having fed her rabbits and caressed her pet lamb, the little girl returned quickly to the house. It was some time before Meeta would venture out alone again. In her eagerness to see the animals she had forgotten fear, but when she caught sight of the cover the recollections of that terrible night came vividly back to her; and all through the beautiful summer, when the trees were white with lichen, where her favourite flowers bloomed, and when sometimes, to her delight, she would find a new pet to take home, nothing would induce her to go out of sight of the house even with her sisters.

In the meantime great preparations were being made at the Park for the coming evening. Squire Haverleigh's residence was a large commodious building surmounted by turrets of grey stone; in the projecting wings on either side of the house were niches wherein stood stone statuettes. A moat surrounded the building. Probably about three centuries ago it was one of the many isolated castles belonging to the barons of that period, as the more antiquated portions of the house showed. In some parts of the neighbourhood similar moats could be seen, though in several cases the buildings which they protected had long since fallen into decay. It was a sweet place at all times, and especially so during the summer months; there were trees in the grounds which were pictures in themselves. Lordly cedars spread their gigantic branches far and wide, the lower ones resting on the mossy grass beneath, and the dark blue-green foliage afforded a refreshing shade from the heat of the sun. The front gardens were laid out with great taste, in fantastic shapes and rainbow colouring. The stone palisades leading down on to the terraces below were carved with pine-apples, grapes, and other fruit, while in and out amongst the stone-work a vine wound its delicate green tendrils. There were some fine hothouses too, and the way to them led through rows of graceful statuary. Some way beyond the grounds was a bridge, cut and carved in the same manner as the palisades. The river in that spot was placid as a lake and bright as a mirror. Many a time had Alice gone on a boating excursion accompanied by Hubert and Ronald, when the lilies were in bloom, their pure white petals shining up from amidst the broad green leaves which rested on the water. And you may imagine how the children counted on going to Squire Haverleigh's, where they were allowed a free range over the beautiful grounds. On this particular morning of the party Ronald was expecting a sailor friend, the son of Mrs. Arlington before mentioned, a youth of about his own age, whom he had invited to join them. Mrs. Arlington's residence was a long way from Haverleigh Park by road, though a much shorter distance by river, and the latter mode of communication was much preferred by the boys when they visited one another. As the hours drew on, Ronald had stood watching for some time expecting his arrival, but no Master Leonard appeared, and he began to fear that something had occurred to prevent his coming. At length, however, he espied the figure he sought advancing slowly up the avenue, and Ronald set off to meet him ; but though Leonard waved his cap he seemed unable to get along.

"Why, what is the matter ? " exclaimed Ronald.

"Oh !" replied the other, " nothing but what can be set right, I hope ; it's all because of that plaguey river. The rushes grow so thickly down by Uncle Arlington's place, that you cannot stick your oars in anywhere for them, and what with the turns and twists I made no way at all, and so resolved to walk. I ran the boat as near the bank as I could, and made a spring for it, but it was farther than I thought, and I went right into the middle of a mud-bank, and scrambled out with a fresh pair of leggings, as you see; my clothes are precious heavy, I can assure you."

Ronald laughed as he replied that " there were plenty of such places about, and the brilliant green might prove very deceptive except to a practised eye. The best rowers would find it difficult at this time of the year, for the water in some places was very low, and the weeds got entangled over the oars, and oftentimes the boat would scrape the bed of the river."

"I disturbed some snipes, too," said Leonard, "that were quietly hiding there, and if I had had a gun I might have been rewarded for my misadventures."

By this time the two boys had reached the house, and Leonard was soon equipped in some dry clothes. They then went out on to the terrace and walked in the grounds.

"This is a pretty place," said Leonard ; "I think, if I lived here, I should be almost tempted to give up knocking about the world, and settle down to a country-life."

"Your friends would only be too pleased if you would settle down as you say," returned Ronald ; for Leonard was Mr. Arlington's god-son, and he had very much wished to make a barrister of him.

"What!" exclaimed Leonard, "would you have me sit in a mewed-up office all day ? It makes me gasp to think of it."

"For my own part," replied Ronald, "I admire the fellows who can quietly put their shoulders to the wheel, and are content to grind, year in and year out, at the same monotonous work, though if I were in the position of many here I should feel inclined to go out to the colonies, and try my hand there ; one might hope for some chance of success with an average amount of ability."

At this moment the luncheon bell rang and the two youths entered the house. While at luncheon Leonard gave a description of the previous day's fishing, which had not turned out any better than his rowing expedition.

After leaving the morning-room they entered the conservatory, a spacious place, which looked delightfully cool with its polished pavement of tesselated tiles. In the centre were some rare and beautiful plants, which though a blaze of brightness, the colours were arranged in perfect harmony, and the exotic fragrance was delicious. On one side, reaching from roof to floor, were ranged masses of virgin cork, and from each interstice different kinds of creepers, maiden-hair and other delicate specimens of fern hung in abundance, while from some rugged rock-work in the midst, veiled by wreaths of hanging moss, a thread of water trickled with a liquid sound, watering the plants in its descent, and every frond was strung with dewdrops.

But the aviary interested Leonard the most. In the centre compartment were canaries ; in another a pair of beautiful little love-birds sat close together on the perch, looking with their bright eyes at Leonard as he came up ; on one side was a solitary Java sparrow, the only survivor of a dozen that a friend had brought for Ronald on one of his homeward voyages, for what with storms and bad weather the poor birds had been unable to brook their hardships and privations.

"I do not see any harm," said Ronald, "in keeping caged birds, though I would never have a lark on principle ; their nature is to soar, and it is miserable to see them beating against the bars of their prison; the song of a caged lark has no beauty in it for me."

At this moment Leonard's attention was attracted to the far end of the conservatory, where there was a handsome white cockatoo, whose head was adorned with a golden tuft. The whole of the time he kept swaying up and down on his stand, and ruffling up his feathers screamed lustily, because the visitors did not take sufficient notice of him. The bird bent his head for Ronald to stroke him, though his eyes looked as if he were not to be trusted should any unlucky stranger venture too near.

Leonard then explained to Ronald that he had a sailor friend staying with him, and he so much wished to come that he would most likely be over with Harold and Algy in the afternoon.

"So much the better," said Ronald in his hearty way, "the more the merrier;" adding that he "should have been sorry if Leonard had not felt at home enough to bring any friend of his to the Park."

Presently from the conservatory they espied Harold with Hector Bellamy and Algy Arlington, coming down the road. Leonard ran off to the gate to meet them.

"So you thought you would come early," he remarked.

"We were tired of loafing about at home with nothing to do," replied Harold, "so thought we could not do better than come on here."

Ronald glanced at Hector, and going up shook hands with him at once, for he was not one to stand upon ceremony. Hector was a tall, slight-made lad, with pale complexion and fine features, and his long hair, raven-black, hung over a lofty and thoughtful brow. He was as delighted with the Park as Leonard could desire, and when Ronald suggested a row on the river, he declared "there was nothing he would like better," and so it was agreed. Harold, who had brought his rod, took himself off at once for an afternoon's trolling. The four boys were soon joined by James.

"May not I go too?" he asked.

"Not this time," replied his brother, "we don't know how long we may be out."

James looked rather cross for the moment, but Algy ran up to him.

"Never mind," said the little fellow, "I don't care, I have brought my ship in the chaise; we can sail her, and that will be better than going in the boat."

In the meantime, Alice, with her basket, was loitering by the river side, searching for some wild hop and other creepers. She had been very busy helping Mildred and Leila during the morning, and there remained nothing to be done now but to arrange the flowers for the table. Presently she came upon the group of boys, who had been hidden from her view by a weeping willow which hung over the water.

Ronald was standing in the boat, guiding her alongside the bank with the oar he held. He looked up suddenly, and seeing Alice exclaimed with his bright smile, "You are just in time, and you always steer, you know."

"I cannot now, I am afraid," replied Alice; "we have not finished decorating."

"You *must* come, Miss Rothesay," added Leonard. "Here, Hector, lend a hand."

Hector took off his cap, and came forward at once.

"Do, Alice," said Ronald. "I am going away very soon, and I must be humoured a bit."

"But what will Mildred and Leila say? I was going to gather some creepers for them, and they will be waiting."

"We sha'n't be gone long," pleaded Ronald, "and my sisters won't mind for once; besides, there are plenty of those trailing things and blue what-you-may-call-it up the river."

Alice laughed, as she replied, "You should have a few lessons in botany, Roy, to teach you the names of plants."

"I shall be delighted," replied Ronald, "provided you are to be the instructress—we must have that in the agreement, mind; and you will come?" he added, "it is lovely up the river."

"It is very lovely," said Alice, as she glanced beyond the shade of the overhanging trees, and on the sunbeams that glinted on the sparkling water. It was too pleasant to be resisted, and presently she was seated on the crimson cushion of the pretty *Kathleen*, and Leonard remarked that "she was in good hands, as there were three sailors to take care of her."

Ronald told him that if Alice liked, she could dispense with their assistance altogether, for she could row as well as any of them.

Hector sprang in last, the boat went all on one side, and they were very nearly being capsized to begin with. Algy screamed out, "Oh, do look at her, she will be over!"

"What a funk you are, Algy!" laughed his brother; "shut up till you see danger, there's a good lad."

"Good-bye, boys, and take care you don't fall in," cried Leonard, as they rowed swiftly away.

James resented the injunction, and called out that "they had better look after themselves."

The afternoon was very hot, a scarcely perceptible breeze stirred the water, and it was the very day for a row. The reflection of the boat and its occupants was clearly mirrored in the stream, while the pale-blue dress Alice wore lent to it a transparent bluish tinge. How delightfully cool it was gliding under the trees, and what sweet music the oars made as they touched the water, and rose sparkling with shining drops, the sides of the boat and banks of the stream reflecting in golden wavering lines each small ripple as it rose and fell. The water was so clear that Alice could see the fish quite plainly hurrying along below, and soon the weeds began to get higher and float on the surface.

"How is Hu, Alice?" asked Ronald; "of course he will come to the party, he will not disappoint us?"

"Oh no!" replied Alice; "he has been quite looking forward to it, and I am sure the anticipation has already done him good."

"I am glad of that," he said; "Hu was always so strong that it seems unnatural for him to be ill."

"He was a brave fellow," said Leonard. "If I had been placed in such a position I fear I should not have stood my ground so well; I should have been like the man in the song, and taken to my heels."

"I am so volatile," laughed Hector.

"Do you see yonder bridge?" said Leonard presently "It is one mass of fossil shells."

"I should like to examine it," said Hector. "I wonder how this place looked some thousands of years back; it must have been sea at one time or another."

"No doubt about that," replied Leonard, "as the shells testify; it is a pity one cannot obtain one of them perfect, but as soon as you make the attempt, the stone is so hard, it shivers into fragments, though you may obtain a piece with plenty of impressions."

At length Alice espied the pretty pale-green leaves of the hop-plant winding in and out of the hedge on the bank, though many of the flowers had formed into seed by this time, as it was getting late in the season. Ronald propelled the boat as near the bank as he could, Leonard stepped ashore, while Hector held the basket, and it was soon filled with creepers and blue spurge.

"Oh, thank you!" exclaimed Alice, "that is just what we wanted."

"Before we return I must show you what I call my garden," said Ronald, and as they went on, the fresh green of the park sloped down to the water's edge, overshadowed by chestnut trees. Under one of them was a rustic seat, some one was reclining on it, with a book in his hand.

"Why, there is Bertrand," said Ronald in a low tone : "keep quiet!" and dipping his oar in the water, he sent a shower of spray over the recumbent figure, who started to his feet at the salute. The boys began to laugh.

"Halloa there!" cried Bertrand. "Roy, I call that too bad."

"How do you know it was I?" he retorted.

"Because you are always up to some game or other. Where are you going?" asked Bertrand.

"Only up the river."

"You had better make haste then," he replied, taking out his watch, "for it is past four."

Ronald whistled, for the company were expected at five, and they were a long way from home.

"Indeed we must go back," said Alice.

"You *shall* see my garden first," answered Ronald ; "we are in for it now, and a few minutes can't make much difference."

As they turned the bend of the river, in the midst of the wide stretch of luxuriant grass was a bed of flowers ; they were arranged most tastefully, and showed that whoever had planted them had an eye for colour, and the little party all declared it very pretty.

"Why, Roy," said Alice, "I did not know you cared so for flowers."

" But I do,". he replied, " and if I had the time I would study their cultivation."

Turning the boat they now quickened their pace, and with a long sweep, and in perfect time, the boys pulled with a will, and the boat shot like an arrow through the stream, which allowed them to rest on their oars for a little breathing time. Swiftly past the waving rushes, past·the bathing-house veiled in ivy, past the tiny rustic bridge, with its one rail and rough-hewn plank, that spanned the creek which led up to the boat-house, where the water shone with a tinge of golden brown from the sun's reflection. Now the banks were smoothly shaven as a lawn, and the whole extent to the park was like one long garden. But somehow the conversation flagged, and scarcely a sound was heard, save the liquid dip of the oars, as they glided rapidly homewards, in the wake of the sun's rays, which turned into bright crimson the red reeds that grew by the river side.

Reaching their destination, Leonard moored the boat to the bank, Ronald shouldered the oars, and presently all were making their way with rapid steps towards the house.

CHAPTER XIX.

THE PARTY.

CHAPTER XIX.

THE PARTY.

A LICE had barely time to dress before the company arrived. She found Mildred and Leila all ready, sitting in the balcony adjoining their bedroom window, the doors of which were thrown open.

"Why, Alice, where have you been?" they both exclaimed. Alice told them, but made rather light of the boating excursion, at the same time handing Mildred the basket.

Both sisters were very pleased with its contents, and Mildred at once ran off to deck the stems of the vases. Down the centre table they were arranged on stands of looking-glass, fringed with flowers and maidenhair fern, the whole having a very pretty effect.

Alice's toilet was soon completed, Leila declaring she looked lovely; and there was no doubt that a good many besides Leila thought so too that night, with her shining waves of fair hair and blooming countenance, her white dress trimmed with pale blue, her only ornament a silver locket and chain. As she and Leila descended the staircase Ronald told them to come into the conservatory, where several of the young people were already assembled, and handing Alice a spray of hot-house flowers and fern, said admiringly, "There, now you are perfection."

"Oh! do give me some, will you, Roy?" asked Leila, and the kind brother complied at once. Ronald would do anything for his sisters.

All now repaired to the grounds, where a large party had already assembled, and the afternoon's amusements commenced with "Les Graces," "French Romps," "Archery," and other pastimes. At length Ronald proposed "Rounders" as a good game, in which all could join, for he, with some others, had been playing cricket. At his suggestion the boys threw down their bats and entered into it with a hearty good will. Several of the little girls were being continually knocked out, and the elder ones declared they ought not to play, but Ronald good-naturedly said he would run for them and get them in again, and uncommonly hot work he found it. No one could get cross or ill-natured when Ronald was by; the only difficulty was, that every one wanted to be on his side, which was impossible.

At length they were so tired they were glad to sit down and rest, after which some of the elder ones betook themselves to a game at lawn tennis. To please the younger children, Ronald with one or two others repaired to the close, and caught Snowberry, who was duly saddled and bridled. How delighted were the children when they knew they were each to have a canter round the field on the beautiful white pony, that stood so quietly arching his graceful neck, and looking round on them as if he knew all about it.

"Meeta shall be first," said Ronald, "because she can ride well." When she was seated he let the bridle go, and she was off at once at a round trot. But with the others he had to hold the pony in, and run by his side. Some of the children screamed, but nevertheless they would not give in, and were not at all willing to get off. "Leave go of the pommel," said Ronald; "sit upright, don't kiss the saddle, and you will soon learn to ride."

Hector, in the meantime, had lugged out the great garden roller, and lifting Meeta on to it, he drew it along, and she had to keep stepping back very quickly as it went, or she would have fallen forwards. What a merry, ringing laugh she had, to be sure, and how fast her pretty little buckled shoes went tripping backwards! But presently the Squire came upon the scene.

"Not so fast, my boy," he said to Hector. "That is rather dangerous; you had better give it up, and have some other game."

Just at that moment the brougham was espied rolling up the avenue towards the house, and away they all ran to receive Hubert, and give him a welcome. The Squire, surrounded by a group of merry faces, was at the entrance to help him out, and presently Hu was lying on the sofa by the deep window, looking out on the gay scene before him. His eye wandered to the hills and on to the varied scenery of the vales below, and away towards the right were visions of the park, and here and there deer might be seen feeding beneath the trees.

As evening drew on, white dresses, and floating hair tied with ribbons of all hues, and pretty little legs encased in silk stockings, fluttered about the wide hall like so many butterflies. Hanging on the shining oak panels were sombre oil paintings in ponderous frames mellowed with age, some of which were of masters and mistresses of the olden time, arrayed in powdered wigs and rich brocaded satins. Beneath these was a marble slab, on which stood a globe of gold and silver fish, some cases of foreign birds, and below shells lay in groups on

the stone floor. Statuettes stood on either side of the
entrance, bearing lamps of stained glass, which threw a
soft radiance on all around.

There was a good deal of whispering among the little
folks, who had heard that there was to be a surprise for
them some time in the evening. The children of the
rector, with Meeta between them, might be seen peering
through the keyholes of locked doors, and doing all in
their power to unravel the mystery. After coffee had
been served, there was a general rush to the field, as the
dance was to be held in the barn, and Ronald said it
would be a warming for the villagers and his father's
men on the ensuing day. All down the gardens Chinese
lanterns were strung from one archway to another that
spanned the path from end to end.

These arches were covered with woodbine and roses,
and went by the name of the Rosery. Glow-worms
began to glimmer here and there in the shade, and the
children declared it was as good as a fairy tale.

"Well said," exclaimed Hector; "we shall be like the
fairies to-night, lit to bed by glow-worms and serenaded
by nightingales."

Presently they all joined in singing the "Wood
Nymph's Call," after which one of the party asked
Ronald if he could give them a song.

"I can't," replied Ronald laughing; "I cannot trust
my voice; I begin all right, and then go off into a key
which I defy you to find on any instrument."

When they reached the barn, the musicians were
already there tuning their violins, and soon there was a
pairing off of partners, and all the little folks as well as
those of larger growth were pirouetting in real earnest.
Meeta's only regret was that Hubert could not be with

them and enjoy it too. Many a time in the evening did she and Alice steal away to describe to Hu what they were doing.

Mrs. Rothesay, in the meantime, was sitting with Mrs. Haverleigh, giving her an account of poor little Meeta's adventures, for she had much wished to hear more particulars. She told Mrs. Rothesay what pleasure it had caused her, that Ronald had been enabled to find the child.

"I know what he is so well," she continued; "he never would have rested if his search had been fruitless, and he would have gone off to sea again thoroughly disheartened."

In the drawing-room the Squire and his friends were discussing politics, and the state of the country in general. When the clock struck nine the Squire ordered the bell to be rung to summon the young people, as many of them came from a distance, and it would be late before supper was ended. The bell was heard by the merry group in the barn, and straightway the fiddlers put up their instruments and made for the kitchen, to regale themselves with substantial refreshments. Ronald started away with Della Cecil in one hand and Meeta in the other, followed by a whole troop at his heels.

"Really, Roy," said Mildred, "I wonder you do not go off with the boys. I should have thought you would be tired of girls' play."

"Not I," said Ronald, "I get plenty of the other company on board ship, rather more sometimes than I care for."

Soon all were assembled in the dining-room. Squire Haverleigh stood at the head of the table, and Mrs.

Haverleigh had, assisted by her husband, come down-stairs that she might see for herself the many happy faces assembled that night. In an instant Ronald was by his mother's side, more pleased than he could express to find her amongst them, and her eyes rested with love and pride on the handsome face of her darling son. He had had scarcely a moment's time to himself, for being a favourite with the children they had clung round him like a swarm of bees all the evening. The eyes of the little company looked somewhat curiously on the well-spread table. In the centre, towering above all, was a huge cake, apparently frosted over with sugar, surrounded by holly, ivy, fruit, and flowers. It was ornamented with flags and various devices, and crowning all was a silver equestrian figure with a sword in his hand. It was to this cake that the eyes of the children were especially directed.

At this moment there was a ring at the hall door, and presently Mr. Cecil entered, the arrival of whom was the signal for all to take their places, which they accordingly did. The good clergyman took his seat by the Squire. When they had done full justice to the dainties set before them, Mr. Cecil rose, and tapping the table to enforce silence he began—" My dear children, it gives me great pleasure to be amongst you to-night, and to see so many merry faces around me on this auspicious occasion. I need not say what delight it has caused me to see our dear little friend, Meeta Rothesay, restored to her home ; not the little pale-faced girl we expected to see, thanks to her kind and careful guardian. Though I did not come here to preach," he continued, " I might say we could all learn a lesson in charity and unselfishness from one who, I do not suppose, ever preached a sermon in his life, or dreamed of such a thing."

The Rev. Theodore here glanced at Ronald, who flushed up and looked away. His mother's eyes glistened with pleasure as she heard him so praised before all there, for there were a good many lady and gentlemen friends of Squire Haverleigh's present besides the children.

"There is no need to be a clergyman to preach sermons," continued Mr. Cecil; "people's daily lives are sermons, and they influence us either for good or ill. And now on this evening, when we are once more restored to each other, we must endeavour to forgive those who have so deeply injured us, however bad and depraved they may be, recollecting that their wickedness is often the growth of gross ignorance; and we who have had the advantages of a good training and education must have mercy on those dark minds upon whom the light has never shone. But I must not dwell any longer on grave subjects, for I see many anxious eyes watching me, and all are in haste, I have no doubt, to see what that gorgeous-looking cake in the centre means. You have all of you, dear children, I daresay, an idea that there is something mysterious about it. It is just this —though it looks exactly like a cake, it is not made of plums or currants or anything of that kind, but it contains presents for you all, with which little piece of news I will close for to-night."

There was great applause at this delightful announcement, and Mildred Haverleigh now came forward with a basket in her hand, and gave tickets round to each one, all of which were numbered. The cake was then placed before Mr. Cecil; Hubert's couch was wheeled up that he might see the presents given out. How they did come crowding up! and what excitement prevailed as each child displayed their gifts ! Many of them were exceedingly pretty; Meeta's was a beautiful model of a barque, which Ronald had carved for her in his leisure hours, with cabins and steps leading down, all complete. But besides this, as a memento of her happy return there was a gold locket and chain from the Squire and Mrs. Haverleigh ; the locket was inscribed with her monogram (set in sapphires), and the delighted little girl was just allowed to look, after which it was carefully put by for her by Mrs. Rothesay. Little Meeta ran and thanked her friends so prettily for their handsome present that they were quite pleased with her. Still it was a question, at that time, whether the barque did not give the child the most pleasure, as she fancied to herself how queenly she would ride over the tiny ripples of the brook at home. When Eva's turn came she had a set of modelling tools, which were just what she had been wishing for, while Mildred and Leila each received a necklet of gold sequins. It was a surprise for Mrs. Haverleigh to find herself amongst the recipients. Her present was a white satin satchel, exquisitely embroidered with Venetian shells.

Ronald's countenance as she looked at him betrayed him as the donor. " My dear boy," she exclaimed, " how beautiful ! " and he was amply repaid as he witnessed the expression of pleasure which kindled her delicate features, for Ronald idolized his mother, and would do anything to promote her happiness.

At length Mr. Cecil drew forth a box tied with blue ribbon. "This box is wonderfully heavy," he remarked. " I wonder what it contains—who is No. 20 ? " Ronald came forward, and Mr. Cecil placed the parcel in his hand, saying with a smile, as if he half suspected what it was, " I have great pleasure in giving you this, my boy."

Presently an exclamation from Ronald made all look round ; the box lay open on his mother's lap, and lo ! there was a handsome gold hunting-watch, on the back of which was inscribed, "To Ronald Haverleigh, in gratitude for his noble and generous conduct." Everybody crowded round, and everybody wanted to see it at once, and it was passed from hand to hand by the admiring throng. Bertrand, who had been rather silent hitherto, turned to his brother, saying, "I should not mind changing places with you to-night, Roy." The fact was, Bertrand could not quite take in all the " palaver about charitable and Christian conduct, and all that sort of thing," as he termed it, but he *could* understand the watch—that was tangible.

Ronald thanked his kind friends the Rothesays again and again for their beautiful present, which he said he should value more than anything else he possessed. Besides the watch there was a pretty little marine pocket compass from Alice. Her lovely face was radiant when her turn came, and she untied her parcel to find a silver frame of delicate filigree work, in the centre of which was a beautifully arranged group of rare seaweeds. She knew in a moment who was the giver. Running up to Ronald she exclaimed, " Oh, Roy, how kind ! it is very beautiful, and I shall prize it most highly."

" I got it on purpose for you," he answered, " on my last voyage."

There was so much laughing at the other end of the room that Ronald and Alice went to see what it was. It seemed that Bertrand's present had turned out to be an ugly little wooden figure with a string attached, which, when pulled, legs and arms stretched out something like the sails of a windmill.

Directly his brother came up, Bertrand exclaimed, "So you are at it again, Ronald! I know it is you I have to thank for this," at the same time contemptuously holding up the ugly little figure.

"Lay it on thick, old fellow," said Ronald, laughing, "my back is broad enough."

But Bertrand found out that he was wrong after all, and that it was the work of Eva and Mildred.

"Hey-day!" said the Squire—"we are forgetting Hu in all this Babel and confusion, and there he lies looking on at other people's gifts and nobody thinks of him.

"The fact is, Hu," said he, in his hearty voice, "it is not that you are left out, but your present was—well, it was rather too large to go into the cake, we could not crowd it in any way."

Hubert declared he had not expected any present, and he hoped the Squire would not think of such a thing. But Squire Haverleigh put an end to that by telling him that he was going to fetch it himself. All eyes turned to the door, wondering what was coming next. Soon there was a tinkle of bells in the distance. Some of the children went up to Ronald, who stood by Hubert's couch with a conscious smile on his face.

"We know you can tell us," they said.

Little Meeta put her hand on his arm—"Do, Roy," she whispered coaxingly, "because I am sure you know."

"Listen," said Ronald, and everybody did listen.

Nearer came the bells! and—yes—there seemed to be the sound of padding footsteps, besides the tread of Squire Haverleigh across the hall; soon the door was flung open, and he appeared, holding by a brass collar and bells a huge dog of St. Bernard! The great fellow looked from one to the other very quietly; he was not at all disturbed, as were some of the children, who were so timid that they ran to take refuge behind the others.

"You need not be afraid," said the Squire, "he is as gentle as Meeta's pet lamb," and then he marched him straight up to where Hubert was lying. "Here, Hu, my boy, is your present! you can easily believe now that we should have found it somewhat difficult to get *him* into the cake."

Hu could not speak for a moment; he had never dreamed of anything like this, and bending down he hid his face on the massive head of the beautiful animal, that crouched towards him as if already owning him as his master. The Squire cleverly turned away the attention of the children to something else, giving Hubert time to recover himself. Presently he took a seat beside him, and chatted away so pleasantly that Hu was soon himself again. But for the moment it brought back so strongly the fate of poor Juno that he was quite unmanned.

"And now, Hu," said his kind host, "that dog must be your companion in all your future rambles, and I will back him against whatever you may choose to take for self-preservation."

"You could not have given me anything I should appreciate more," replied Hubert warmly; "he is a splendid fellow, and I shall certainly take him with me wherever I go, and shall be proud to do so."

Meeta bounded across the room to caress the dog.

"What a child she is for animals!" said the Squire to Mr. Rothesay.

"She is," he replied, "and chatters away to them as if they understood all that she says."

"They comprehend more than we give them credit for, I'll warrant," answered the Squire.

"What shall you call him, Hu ? " said Meeta.

" I really cannot tell just at this moment, you must set your busy brain to work to think of a name, and if it is not too magnificent and unpronounceable—well ! I will keep to it."

Just at this moment a servant came to the door, saying that Mr. Rothesay's chaise was waiting, which was a signal to depart. But before Meeta could obey the summons, the Squire caught her up, giving her a kiss and warm embrace as he bade her good-night, and then he watched her light figure as she ran airily away, arrayed in her simple white dress, her long golden hair fastened with a broad band of pale blue, her only orna-ment. Little Meeta had never breathed any but the pure atmosphere of the sweet home life : she had not the conscious air of many of the little would-be women of the present day, who trip from one to another to be flattered and admired. It was her sincere, true, and childlike nature, the sweetness of expression that lit up her blooming countenance, of which she was wholly unconscious, wherein lay her chief charm, and it was this which so delighted the Squire as he raised the un-suspecting child in his arms to embrace her.

When little Meeta went up to Mrs. Haverleigh to thank her for the happy evening she had spent, she drew the child to her side, and told her how glad she had been to welcome her home safe and well. " But this must not be the last time, my dear Meeta," continued Mrs. Haverleigh ; " I hope that you and your sisters will come for a long day here before my Ronald goes to sea."

" I wish he need not go ! " said Meeta, with a quiver in her voice.

" So do I, love, but every one must have their troubles,

and I hope he will be soon back again to me this time. Poor Jamie will be delighted to have you, for he will be lonely when dear Ronald is gone, and you could trot over here easily on your donkey if your mother is willing for you to do so."

Once more thanking Mrs. Haverleigh for her kindness, the little girl ran off to Alice, who wrapped her in her scarlet cloak, packing her up snugly in the back of the chaise.

"Who is going in the brougham with Hu?" asked Mr. Rothesay.

" I am," replied Eva, "and Alice and Roger are going to walk home with some of the others."

"Come, Alice," said Roger from below, as he stood leaning over the balustrade. She soon appeared with a light plaid thrown over her shoulders.

Ronald was waiting too, and talking to Roger when Alice joined them.

The harvest moon was rising large and beautiful above the clouds which lay here and there like crystal islands, and her pale blue-green light fell in lines like hoar-frost across the grass.

The vehicles were before them as they passed out of the park gates, and the wheels could be heard distinctly as they came every now and then to a bend in the road.

The three were very silent; as they went along it seemed as if they had left all the fun and merriment behind them.

" You must come and spend a day with us, Roy," said Alice at length, " before you go."

"I shall certainly call," he replied, "but I cannot promise to stay, much as I should like it,—my mother must have my time now."

"When do you sail?" asked Roger.

"I can't say precisely, I expect a summons to Plymouth in a few days at latest."

"You will not be away so long this time?" said Roger.

But instead of answering lightly as he usually did, Ronald replied, after a moment's pause, "I hope I may not, for my mother's sake, but a sailor's life is very uncertain. It is a disappointment to me to have had so little opportunity of coming to the farm, we haven't had one quiet evening since I came ashore."

Roger felt there was a shadow of regret somewhere, ut he did not know how to answer him.

Alice was wishing to express a kind word of sympathy, and to tell him that by the household of Fern Glen he would never be forgotten, but would be in their prayers always; but her brother's presence and her shy nature kept her back. To Ronald, if she had only known, the knowledge of her feeling for him would have been doubly precious just now on the eve of his departure; he longed for the sweet sympathy of a soul akin, he was afraid to speak to his mother, or to hint at a fear for which he could not account.

Presently they were joined by some of the others, and so the word which might have changed the whole tenor of that evening's conversation was never spoken.

"Give my love to them all," said Ronald, as they parted at the gate, " and tell them I shall come and wish them good-bye."

Alice found herself almost mechanically bidding him good-night, and she stood and watched him as he sped along the moon-lit road, and his lithe form and bright face came back to her many a time afterwards.

CHAPTER XX.

THANKSGIVING SUNDAY.

THE whole of the next morning Alice was watching for Ronald, as she had promised herself another opportunity of speaking, and until late on the previous night she had been thinking over all that she wished to say. It was getting late in the afternoon, and Meeta was just going to help Eva look after some little guinea-chicks which had strayed into the garden, when she caught sight of Mildred Haverleigh driving across the field, and started off at once to welcome her.

"Are Eva and Alice at home?" asked Mildred.

"Oh, yes, do come in," said little Meeta, "and I will run and tell them."

Presently they were all seated in the drawing-room.

"I have come with a message from Ronald," said Mildred. "He is so sorry, but he received a summons this morning, and he had to go away at once, without bidding any of his friends good-bye. He will not really be off, I expect, just yet," explained Mildred, "but all are obliged to be on board, as the ship may set sail any day." Then turning to Alice, she asked gaily when the long-talked-of scheme of studying German together was to commence, for hitherto something had always occurred to prevent. Leila was going to school directly, and then she would be glad of Alice for a companion.

But Alice could not converse in the light vein of Mildred Haverleigh—she felt how differently Ronald's departure would have affected them had he been their brother.

Mildred could see that something was amiss, and that no one seemed much in talking humour, so after a few more commonplace remarks she wished her friends " good morning," telling Leila on her return that they were so slow at Fern Glen and that she could make nothing of Alice, who would scarcely speak, and ended by declaring that Ronald might go for himself the next time.

Directly Mildred Haverleigh had left the house, Alice ran away to hide her grief, perhaps the most remorseful in all her girlish experience. " The opportunity has passed—he will never know now!" was her sad reflection. But there was one thought that gave her comfort, though that did not come until long afterwards, that he would know one day; not in this world perhaps, but in the higher light of that other existence, she would find her lost opportunity again, and all that she would so gladly have said, but had not been able to express, he would discern in one clear single glance.

As Meeta was sad on account of Ronald's sudden departure, Mrs. Rothesay called at the rectory to ask if the two children might come and spend the afternoon with her little daughter. Mr. Cecil readily consented, and soon the three little folks were running towards the house. Alice was standing ready to welcome them ; she was apparently just the same, and her absence had been little noticed—she so frequently went away to study or to paint ; now, putting aside all thoughts of self, she was as ever ready to give pleasure to others.

She now suggested a ride in Meeta's pretty little

goat-chaise, which was hailed with exclamations of delight by the two children. Nanny was more timid than was Kitty, and more difficult to be caught; but Alice took her by her horns, while Kitty followed closely after, bending her head to be caressed. When the goats were harnessed, Kenelm took the gaily-coloured reins and mounted in front, while Della sat behind, Alice and Meeta walking by their side.

After they had taken several turns down the road to the gate and back again, Meeta took her little friends to see the yellow ducklings which had just been hatched, and which were snugly ensconced in a basket in the kitchen. Then they visited the shed, wherein was a lovely Alderney calf; it was not at all afraid of the visitors, but allowed them to stroke and caress it. After the children had roamed about the farm and seen the cows milked, it was tea-time; and the three ran off to get Meeta's little pink tea-service, which she was always allowed to use when she had company of her own. Nettie set the tray for them, arranging cake and pastry on the miniature dishes, and this was a never-failing source of pleasure to the children.

Towards sundown the children's nurse came to take them home, Meeta ran with them to the gate; the setting sun turned the clouds into burnished gold, lighting up the small receding figures, and the gleaners approaching with their ripe bronzed sheaves in one red glow of light.

The following day was Sunday, and when Meeta came downstairs, she looked for the pretty knot of flowers that she knew would be awaiting her. For on that day it was Mrs. Rothesay's invariable custom to place flowers by each plate on the breakfast table, and this morning little Meeta thought that they had never looked so sweet and fresh before. Each small spray was differently arranged, and carnations and roses were interspersed with lemon-plant and thyme.

After breakfast all, with the exception of Hubert and Alice, wended their way over the fields to church. The morning was somewhat chilly, but the gay hearts of the little company did not feel it.

"I call this my Thanksgiving Sunday, mother," said Meeta, with a happy smile, looking up into her face as she held her tightly by the hand.

"It is to all of us, my child," replied her mother, "and we must especially remember dear Ronald too to-day, and all 'those who go down to the sea in ships.'"

"I do, mamma," said the child, a pensive look coming into her face—"I never forget our sailors, for Roy's sake."

How the villagers turned and looked after Meeta as she tripped by her mother's side! She had become quite a little heroine in the neighbourhood, and as she neared the rectory Kenelm and Della Cecil ran out to welcome her. The church looked very pretty, with its simple decorations of flowers and fruit, while small sheaves of wheat stood in the deep Gothic windows.

As it was the harvest festival, the remarks of the rector naturally turned upon the ingathering of the grain. It was a glad sermon, a sermon of thanksgiving, to which the little assembly listened with such interest on that memorable Sunday morning. To one small company who sat in the large square pew there were hidden depths in the remarks which could only be fathomed by those who had been called to similar trial. Sorrow and sore perplexity had vanished in the brightness of blessing restored and in joyful reunion.

The good rector reverted first to the slumbering seed, which awoke at length into tremulous life in the cold and darkness beneath the ground, its growth how slow but sure—all good things required patient waiting for. Dark clouds swept over it, rain and wind beat upon it, but the blessing, though hidden from sight, was surely there, and would blossom at last into the full, ripe, golden grain. So against our earthly tenement, storms of sorrow beat and clouds of affliction hover, but nothing can extinguish the immortal spark which lies hid beneath the frail shroud of humanity—as tender as the seed, yet capable of a growth far deeper, far higher, a reaching upwards into a sphere now as much beyond it as the tiny seed differs from the golden-eared grain. And as he spoke, the mist and perplexities obscuring mortal vision lessened in the unfolding of a heavenly plan, throughout which the finger of a Deity might be traced. Mystery, obscurity there must be, but the eye of faith with keen gaze pierced through all doubt, and in spiritual rapture rose to rove free and unfettered in the very dwelling of that Power who " holds the waters in the hollow of His hand."

" So," said the preacher, " as the revolving seasons come, each with its Divine lesson of faith and patience, let ours not be the only voices mute, but let us all unite in the glad anthem of praise to Him from whom we receive so many precious gifts."

And joyously the notes arose from the full hearts of those assembled there, young voices sounding shrill and high above the rest.

CHAPTER XXI.

RONALD'S LETTER.

CHAPTER XXI.

RONALD'S LETTER.

IT was some weeks now since Ronald's departure, when one or two warm days in the waning year indicated that summer was loth to depart. Hubert, who was still delicate, would lie on the lawn in the rich October sunshine, while Alice sat by his side busily finishing a painting in water colours.

It was a sweet peaceful scene; rich tints of russet brown studded the hedgerows, and hues of deep crimson peeped from the thick woods in the distance. Day by day the birds were becoming more silent, the trees were beginning to cast their leaves, the very winds as they sighed past seemed to whisper that winter would soon be here.

The whisper could be heard too in the clear twitter and trilling note of a small robin, that perched on the iron railings near at hand. All this, unnoticed by ears less refined, had a saddening effect upon Alice; the still voice of nature seemed to address itself to her in particular, that the fading of the leaf and floweret was not all, that there was something more. She knew that the mellow autumn sunshine would soon be hidden behind the ice-veil of snow clouds, and that the bright landscape would become grey and barren in the drenching rains of winter.

She knew that the soft summer winds would change into wailing gusts, which would find their way into the house and moan in at the doors and windows. In her mind's eye she could see afar off one vast heaving sea, with great surging billows tipped with angry crests— and rocking up and down a solitary vessel scudding before the blast with streaming pennon. She could hear the stormy petrels scream as they hung in the air and flew above, beneath. Would he return ? or were those eerie-like birds uttering a dirge for him too as they had for so many who had gone away and never come back !

But as time drew on from autumn into winter, there were blazing fires and Christmas trees, and glad children's voices were heard up and down the staircase as they shouted and played at the old games of " blind man's buff " and " hide and seek."

Sometimes as a great treat there were fireworks to conclude the evening's entertainment—squibs that darted about with a hiss and a splutter, and St. Catherine's wheels, some of which refused to go off, and appeared in the darkness like a great red eye being frantically whirled round and round ; but it was of no use. This did not trouble the children, however, who hailed everyone with a fresh shout of delight. And Alice, the children's friend, surrounded by the merry, happy bairns, forgot her once sad musing in the autumntide. Then there were the church decorations, which devolved almost wholly on Eva and herself, for Mildred and Leila lived too far off to come often in the winter.

But when the frost began to show his silver rime, and the ground was white and hard, then was the time for sleigh and skates ! There is nothing more beautiful than a bright frosty morning in the country. All along the hedgerows, where busy spiders had spun their webs, seemed covered with spangled gossamers, every blade of grass was strewn as it were with diamonds, and the trees, glittering with crystal drops, stood out white and frosted against a background of limpid blue.

Then Meeta with her sisters would hie to the ice-fields, where ragged urchins were busily sweeping away the snow with their brooms. But there were times when the snow would collect in great drifts, and block up all egress from Fern Glen even to the church; but, notwithstanding, the Rothesays managed to spend the winter months quite merrily. Very frequently they would prepare some pretty little drama to enliven the evenings, to which the older members of the household would be formally invited.

At length the first faint signs of spring began to appear in the silver of the snowdrop and the gold of the crocus. The air was fragrant with the delicate scent of the almond tree, and as time went on, the sun lit up the laburnum's golden chain, and the lilacs arrayed themselves in their mauve blossoms. The windows are thrown open to the breeze, and the peacock can be seen on the low sill, sunning his gorgeous plumage. Hubert at last, to the great joy of the household, was able to go about the farm as he used to do, while little Meeta, glad as the birds, was running about the fields again, rejoicing with a new gladness in the fresh young green, and in the warble of the lark as he rose at her feet.

Spring too awoke in the heart of Alice, and already she was beginning to look forward to the time when there should be news of Ronald's voyage homewards. It was one sunny afternoon, a quiet day like many another, on which the sun had risen and set, but it proved a memorable one to Alice. Roger, who had just returned from the town with the letters, gave them to Meeta, and presently she was running lightly up the staircase, calling to Alice that there was one for her, a letter bearing a foreign postmark, and directed in a bold, decided handwriting, the characters of which brought the colour into the cheek of Alice, and made her hand tremble. It was as if in that moment some dreamy thought that had dwelt dormant in her mind, of which she had been almost unconscious, had now confronted her, and with a keen questioning was reading her inmost soul. Little Meeta stood by, looking up wonderingly at her flushed countenance. But she was not a child to ask questions, she knew that the letter was a foreign one, and at last settled it in her own mind that it might have come from Ronald, but why Alice should look so disturbed she could not determine. Left to herself, Alice quickly turned to her room. The seal bore the initials she knew so well, and hesitatingly she opened and read the first letter Ronald had ever addressed to her.

It commenced with graphic descriptions of the various countries he had seen, of hot dry climates, of his visits to scenes that held a sacred place in classic lore, and then the writer broke off abruptly, and began in quite a different strain. He spoke of his longing to be at home again, and that there was nothing like the old country after all ; but there was something else besides home that had attractions for him now, something that made him, as he expressed himself, oftentimes absent and forgetful of what he was about, that caused him to be more anxious to rise in his profession than he had ever been before.

"And now, Alice," he continued, "this that I am

about to tell you has to do with yourself alone, and I wish it was in me to express what I have to say ; on any other occasion I have had but to face an obstacle to become master of it, but this is a different thing altogether. Do you remember that last evening before I set sail—the row on the river, and the moonlight walk ? I cannot tell you how many times I have dwelt upon every incident of that day, when miles on miles of sea have been rolling between us, and I have wondered if you have given me a thought sometimes, when the wind blows, and keeps you awake with its moaning on stormy nights. When I was far away I began to see you as you are, Alice, and to realize how true and good and beautiful you were. For as I look back, pictures of the old times are ever recurring to my mental vision ; of the nutting parties and of the long walks violeting in springtime, but it is your figure alone that stands out in bold relief from them all, the rest is in shadow and is dim and obscure. We quit this port in a day or so at latest, as affairs begin to wear a more peaceful aspect, and no one will be more glad than I when the *Hero* turns her head homewards. I find myself often wondering whether it will be you who will first come out to meet me when I set foot once again on the dear old farm. Forgive me if I am wrong in daring to say so much—if it should be so this time of my return will be the happiest I have ever experienced. 'Dinna forget'

"Yours,

"RONALD HAVERLEIGH."

This letter was the step that separated Alice for ever from her world of girlish dreams and hazy imaginations. Like a bud that expands and looks up to welcome the

15

first sunbeam, so did the heart of Alice rise and go out to meet this first love. Sweet as was her face before, it was beautiful now ; her eyes shone with a new light, and her colour came and went with a conscious flush as she lingered before the glass to smooth her hair ere she descended to the dining-room, and her light footstep seemed lighter as she quickly went downstairs. The family had already assembled at tea, and Alice had plenty of inquiries as to her letter, for foreign missives were not so plentiful in the small village of Fern Glen.

She told them that the letter was from Ronald, and that he expected to be home again very soon now.

" Tell us something else about it, Alice," said Roger, " for Ronald is a first-rate scribe."

" I cannot tell you more," replied Alice, absently, " except that he is looking forward to his return."

" You said that before," answered Roger, laughing. " It must be a precious short epistle if that is all—I thought sailors were generally celebrated for spinning long yarns, and I did not know before that Roy was any exception to the rule."

At this juncture Mrs. Rothesay looked towards Alice enquiringly, but her quiet demeanour stayed the question that rose to her lips.

" Come, Roger," said Hubert at last, as the former continued his joking remarks, much to Alice's discomfiture, " leave off chaffing : I thought you were going down into the fields about that piece of machinery of yours, and it will be dark presently."

Alice was very glad when the meal was over, and she made haste to escape before any one had time to notice her, or to make more enquiries. Presently she

was hastening down the garden, and taking a secluded
path she passed through a small iron gate which led
toward the wood. In this retired place, hedged in as it
was on every side, there was little likelihood of her
being disturbed. Life had suddenly grown very sweet
and bright; since the afternoon all around her seemed
to wear an altered aspect. Even the sigh of the leaves
overhead had a new meaning, and whispers of the happy
days to come were sounding in the soft winds which
floated by. The flowers that crowded round her path
looked up with fairer faces as she passed, a purple mist
hung over the clover fields near at hand, and the air
was laden with their fragrance. The fresh pure beauty
of that exquisite summer evening came back to Alice in
after days, bringing with it something of the feeling that
glowed in her heart at that time, of a happiness too
deep for utterance, which rose to find its answering re-
sponse in the calm unfathomed depths that reigned above
and around her. She would not tell any one about the
letter yet, she knew she could no longer speak Ronald's
name like that of any one else, without feeling that every
one must discover her secret, and so she determined to
remain silent. She could not bear for him to be made
the subject of her brother's jesting remarks—Ronald's
name was sacred to her now, he was her hero, her idol.
Even to a casual observer the handsome young sailor
would excite interest—how much more so, then, in the
eyes of the lovely country maiden, on whom he had
bestowed the first affection of his true heart!

CHAPTER XXII.

JAMES AND MEETA.

CHAPTER XXII.

JAMES AND MEETA.

NOW that the warmer weather had returned, early one fine morning Jamie Haverleigh set off on his pony to call on Meeta. The ride in the morning air had brought the colour back to his face, and had lent for a time at least a new lustre to his eyes, and little Meeta declared that Jamie had become quite well and strong again.

"I am so glad that you have come," said Meeta, "for now I can show you my new rabbits, and the kids—we have two such pretty ones."

When the children had inspected the animals they turned towards the house, and Jamie begged Mrs. Rothesay to allow Meeta to accompany him to the Park and spend the day there. Mrs. Rothesay at length consented, but only on condition that the little girl returned by daylight. Jamie explained that she need have no anxiety about Meeta, as the boy Linton should take charge of her all the way home.

Mrs. Rothesay could never overcome her fears lest by any chance the child should be out alone, so that Meeta did not visit so often as she otherwise would have done; but if there came an invitation when Hubert was by, he always put in a voice, saying he would manage to call for her in the evening, so that if it had not been for her kind big brother Hu, the little girl would have had to forego many of the pleasures in which she so much delighted.

At length, little Meeta on her donkey, and Jamie on his pretty pony "Jaspar," were trotting down the lane towards the Park.

When the children reached their destination, their first care was to see that their animals had a good supply of corn, after which they ran off to the house, quite ready for the early dinner provided for them, for both children were somewhat tired, and glad to rest after their ride.

When they had dined, Jamie made off to the pantry to beg some bread and biscuit, and he reached down from the shelf a small jar of virgin honey from his own bees, which he gave to Meeta; he also provided himself with some spoons and plates, all of which he packed into a small knapsack, which he slung across his shoulder.

"Where are you going, Jamie?" asked little Meeta. He told her to wait awhile and she would soon know; and Meeta was obliged to be satisfied.

Retracing their steps through the hall, and turning into a long passage which led to the back of the house, they passed through a Gothic doorway at the end, and then the children were off for an afternoon's ramble. At the end of the terrace on which they were walking was a flight of steps, which led to the corridor of statuary before mentioned. Some of the figures were of clear Parian marble, and at one of these Meeta always stayed to look when she passed that way.

It was a statue representing "Charity." The proportions of the figure though slender were perfectly rounded, the wings were folded about it, and the long hair fell in waves on the shoulders. The head was partially raised as if in prayer, the white hands were extended, and the whole attitude of the figure combined the ideas of a sheltering care as well as of a holy pity and tenderness.

None but an inspired hand could have designed that face! It was not only that the features were faultless in symmetry, but so divine was the expression, so intense the love that beamed from the large intent orbs, and the exquisitely-moulded mouth, that it must have rivetted the eye of the most careless beholder.

To little Meeta, as she stood there with rapt gaze, a halo of light appeared to play around the head, the features seemed to relax, and the whole face smiled.

Behind the statue was a window of amber glass, and the sun's rays through it aided the fancy. To Meeta it was no longer a cold solid block of marble, but a thing of life, shedding on all around the influence of its pure loveliness, and over the heart of the child stole the sense of a great happiness, the meaning of which at that time she could scarcely explain to herself, but it was a dawning perception, though faint and indistinct, that there existed in reality over the whole universe just such a guardian care and love, of which this was but the type.

Forgetting where she was, Meeta was quite startled when Jamie, tired of waiting, called her to come and see his favourite, which was at the other end of the building. It was a small statuette in Parian marble of an athletic cricketer in the act of batting.

At length, opening a door of stained glass, the children passed out into one of the covered walks, shaded by petunias, passion flowers, and other climbing plants,

while tufts of powdered auriculas with their velvet leaves lined the path.

Leaving the walk, they came to some rustic cages where silver-pheasants, pigeons, and other birds were kept. Going into one of these, Jamie pointed to the fantails. "Those belong to Mildred," he said, "and one is quite tame." Directly the bird saw him it flew down and alighted on his arm. "This is Sneeflocken," said Jamie; "is it not a beauty? The one there in the corner is Blanche, but she is too timid to let you touch her."

After feeding the birds, Jamie gave Meeta some biscuit for the water-fowl, for she loved to watch them come sailing up, arching their graceful necks, expectant of the crumbs, which were never forgotten. He then told her that he was about to take her to a spot where she had not been before, and to which he never asked any but his special friends. That James should regard her as his particular little friend and confidante gave to Meeta no small degree of pride and pleasure, as he was some few years older than herself, and with a heightened colour and expectant face she tripped on gaily by his side.

The country in this part was completely embosomed in trees, and it was as solitary as lovely. Not a vestige of any habitation or any living being could be seen, nothing but thick woods on every side, deep shades and golden glints of light, and little Meeta was tasting her fill of happiness; she revelled in the sweet pictures that Nature was ever spreading before her. She understood the whispering voices that breathed upwards from the rich vegetation around. Meeta knew her in all her moods, and loved her in each. To find out the secret habitation of some shy floweret or fern-like moss, to rejoice in the sweet odours that rose from the fresh untrodden earth, was to the child a never-failing source of joy.

Presently Jamie exclaimed, " Look, Meeta, here it is," and in the midst of a group of trees near the water's edge, standing a little back from the rest, was a weeping ash, and the long graceful branches swept the ground. Putting them aside, Jamie made Meeta creep underneath, and she found herself in quite a large space. In one corner was a box containing a flint, a jack-knife, and some other tools. Jamie had also made a fire-place of bricks. In the centre stood a table, formed of a felled tree, surrounded by two or three roughly-hewn seats. Close by the entrance was a rockery of large stones of pink and white spa, interspersed with ferns, which grew luxuriantly, as the shade exactly suited them. Meeta's thorough appreciation and evident delight pleased Jamie.

" I knew you would like it," he said : " is it not a famous hiding-place ? "

" It is so good a one," said little Meeta, looking rather grave, " that if I were to come alone, I fear I should never find my way back,—why," said the child, " there is not even the tiniest pathway to guide you."

" I will show you my landmarks," replied Jamie; " you would never guess unless you knew," and then he pointed out some slips of wood painted white, which were placed at intervals the whole way, but they were hidden by the overhanging foliage. " Jack made the seats and rockery," continued Jamie, " and I constructed the fireplace." Then off he went to get some twigs to kindle a fire, that he might show Meeta how well it burned.

While Jamie was lighting it, Meeta gathered some leaves for dishes, and out of Jamie's knapsack she took the rest of the cake and biscuit left from feeding the waterfowl, and set it out on the table with her little jar of honey. Then the children seated themselves on the mossy turf to enjoy it, and little Meeta thought no feast had ever tasted so delicious as that beneath the weeping ash.

When they had finished, Meeta still remained looking silently out from the bowery cave. Below her the water was dark and turbid, but farther down, as it turned upon its way, it became bright and clear again; the air was so still that not a ripple disturbed the surface, and there was no sound except now and then, when a fish leapt up with a splash, leaving for the moment a silver track behind, and it seemed to Meeta as she looked down intently where the shadows of the foliage lay deepest, that she too was moving away as rapidly as the swift silent current beneath. But her companion was restless, and wished to be going, and he wondered what she could see to interest her so much. The charms of such sweet spots as these by which he was perpetually surrounded he was not so quick to see and feel as the thoughtful child by his side. Very reluctantly Meeta rose, and they turned homewards. As they went on, a covey of birds which were hiding in the long grass rose at their feet with a whirring sound, and quite startled the children. "Look, Meeta!" said Jamie presently, pointing to a field near at hand, "do you see those flocks of birds wheeling over the grain? They are watching their opportunity when the boy is away to help themselves. I remember the last time Roy was at home, he went out at that gate with his gun over his shoulder to scare them away." When the children reached the house, Mrs. Merton, who was sitting at the open window with her needlework, looked up.

"Make haste, dears," she said, "for tea is awaiting you."

They then went down several wide steps to a pretty room draped in blue. In one of the long narrow windows lay a large tortoiseshell cat curled round fast asleep. But Mab, the little pug dog, was wide awake enough, and capered about Meeta as if she were delighted to make her acquaintance. The two children then took their seats by the table, which was drawn up facing the windows. In the centre was a dish of fine peaches and nectarines gathered from the hothouse.

Presently the clatter of tea-spoons, and the fragrant odour of hot tea-cake, awoke the cat, which immediately leaped from her soft seat behind the curtains, and she began to purr and pat Meeta's hand with her paw as a gentle reminder.

Just as tea was over Linton came to tell Meeta that her brother had called to take her home. Since he had been at the Park, Linton had grown a tall lad, and had proved himself worthy of the kind interest Ronald had taken in him.

When Meeta was dressed, the boy with Neddy was awaiting her, and presently she and Hu had started, while Jamie ran down the avenue after Meeta to see her fairly off, but he called her back so often on the pretence that he had forgotten something he wished to say, that it would be a question after all whether the night or little Meeta would be the first to arrive at Fern Glen.

Hubert as usual was followed by his faithful dog, which quite filled poor old Ben's place in watching and attending on Meeta, of whom he was very fond. He would carry her basket and accompany her on all her rambles.

As they went on, the bats began to come out of their hiding-places and sail backwards and forwards, and every now and then a solitary bird would fly across the sky to its nest in the trees, while not far away, when the warble of the thrush grew fainter they could hear the plaintive clear note of the nightingale mingling with the sing-song note of another pert little bird, which Meeta declared ought to have been in bed long ago.

When they reached the rustic bridge, the long rushes, whose plumy heads waved with so bright a gloss in the daytime, now looked as black as night, and little Meeta almost trembled, for not a leaf rustled nor a blade of grass stirred.

" Oh, Hu ! " she cried, " how glad I am you are with me, for everything looks so strange at night."

" You must not be frightened at shadows, dear," answered her brother, " there is nothing to hurt you now; the birds are not afraid, though they go to roost on the topmost branches of the trees, with their heads under their wings and perched on one foot ; if they troubled themselves with fears of falling, what a miserable time they would have of it."

" I know that," said the child, " but ever since that dreadful time in the cover I have been afraid."

"People older than you are, dear, would have been afraid," replied Hubert, " but you never need fear that man again ; the shadows are all gone, Meeta, as far as our people at Fern Glen are concerned, and as we have been preserved in the past, we ought to have hope for the time to come. Of course in this life we cannot expect things to go just as we would wish, yet whatever transpires it is right to face it bravely." Thus, with a determination to meet the unknown future in a manly and resolute spirit, Hubert and his little sister took their way homewards.

CHAPTER XXIII.

THE TRIAL.

CHAPTER XXIII.

THE TRIAL.

ALL this time the man Winter had been in prison awaiting his trial, which was now at hand. For some cause it had been postponed, and he was consequently still kept in suspense as to what his sentence might be.

He had hoped to the last that by some good fortune he would get off with a lighter punishment than others expected to whom he had sometimes appealed for their opinion.

The Rev. Theodore Cecil had visited him several times, and had endeavoured to awaken in him something like a right state of mind, but all the feeling he showed on the subject was that he had been traced and captured just as he had begun to take heart and believe that he should get clear of the country, and escape the punishment he deserved.

The morning of the trial, however, arrived ; the sun streamed in hot and fierce through the windows of the court, where the people were assembled to hear the case. The principal witnesses were Hubert Rothesay and Meeta, Detective Adams and Jack Linton.

Mr. and Mrs. Rothesay would have spared their little daughter the painful ordeal of attending if they could have done so, but that was not possible.

Hubert did all he could to reassure her, promising that he would be close by all the time. When the prisoners appeared at the bar, there was a visible sensation amongst the assembly, and occasional hisses could be heard, and cries of " Order " had to be repeated more than once before silence was restored.

Squire Haverleigh, too, was there, taking a deep interest in all that passed, his tall fine figure standing out conspicuously from the crowd. Meeta, with her golden hair and sweet little face, excited universal interest and admiration. When she gave her evidence Hubert was permitted to stand by her, and she kept her hand fast locked in his, and her face steadily averted from the prisoners. But everybody spoke so kindly and told her not to be afraid, that the little girl began to take courage, and to answer the questions put to her clearly and intelligibly.

Winter watched the evidence as the trial proceeded from beneath his shaggy brows. He saw as he looked around that there was not a countenance that expressed the slightest pity for him, but he did not expect it, for what had he to do with pity ? He wished that something might happen that he could get off and slink away from that sea of heads, that seemed all eyes, looking him through and through.

He saw that there was no hope for him, and his countenance fell, and he began to lean heavily on the rail before which he stood. The proofs were all too clear to leave the slightest doubt of his guilt, and the case was so decidedly against the prisoners that when put to the jury, the foreman arose and said " there was not any need for them to quit the court, as they were all unanimous," and the fate of the prisoners was decided in one short word, "Guilty."

The judge then remarked that the opinion of the jury was quite in unison with his own, and he sentenced both the prisoners to penal servitude, one for five years and the other for life. " The whole scheme," he continued, " was from beginning to end a most cruel and heartless one, and as one crime generally grows out of another, so it had done in the present case, and the prisoner might be thankful that the life of his victim was spared, otherwise he would have suffered the utmost penalty of the law. The world," continued the judge, " is just in a sense, though it sometimes commits great errors, but when there has been an injury done to the weak and innocent, its voice is always heard in loud condemnation of the offender, and no punishment is deemed too severe for him who can look on the helpless object of his cruelty with pitiless eye." His lordship also alluded to the absence of one from the court that day who should have appeared amongst the chief witnesses.

When the result of the trial became known to the villagers, they stood about excitedly discussing the sentence, and expressed themselves as heartily glad that the man was clear of the neighbourhood, for to the last many had feared that something might occur to prevent the law taking its course, having almost a superstitious dread of him, and scarcely believing that bolts and bars were strong enough to save them from his evil machinations.

But their fears were now at an and. The dark clouds that had for a space cast their shadows over the clear skies of the happy village had passed away.

It had been a sad experience for all, involving the innocent as much as the guilty, but by its severe teaching the discontent that had reigned in the breasts of many was effectually uprooted, and it would take a great deal (if indeed anything ever could) to shake the implicit trust and good faith which existed towards the household of Fern Glen, and as time went on this feeling rather ripened than diminished.

Now there were grateful hearts, hearts that truly appreciated the worth and goodness of their kind and lenient landlord. The story of Winter's crime and little Meeta's voyage, so desolate and alone, to the New World, was often related by the fireside to many an interested little group, as they crowded round in the dusky twilight, and though a tale many times told it never palled upon the young hearers.

The effect of the good rector's teaching had everywhere become manifest—the people were governed by other motives than had once dictated their actions. The love of right for its own sake, a cheerful content and a spirit of honesty and frankness, had at last with their blessed influence pervaded the atmosphere of every peaceful cottage home.

CHAPTER XXIV.

CONCLUSION.

CHAPTER XXIV.

CONCLUSION.

SOME few years have passed away since the events took place which have been related. The old grey church is much the same, except that the shining ivy hangs about it more thickly than ever ; the autumn sunlight streams in through the narrow stained glass casements, and their rich colours of purple and ruby rest on the golden head of Meeta, as she wanders softly up and down the aisle, a thoughtful child. The church doors are always open, thus many a weary wayfarer can come to rest beneath its peaceful shadow and go on their journey refreshed and solemnized. Meeta loved to sit in that quiet grave-yard, with the clear cloudless sky above her, and to hear the soft wind playing in the fluttering foliage of the aspens, and in all nature she would see the ever-loving care of her Father in heaven.

It was a good thing in many respects that Meeta was only a girl; if she had been a studious boy she would probably in her thirst for knowledge have come in contact with the hard abstruse reasoning of some sceptical professor who believed that the God of Nature worked by laws rigid and inexorable, and that she was but an atom in its vast realms unnoticed and uncared for ! But happily for little Meeta she knew of no such theories, she was but a simple little maiden of twelve summers, a child of the woods and of the sunshine.

Every fragile leaf and every tender blade of grass or pink-tipped daisy passed over by more careless eyes had a lesson for her. Unconsciously she had obeyed the teachings of Scripture to the very letter, and received the truths of the kingdom as a little child in perfect faith and confidence.

And perhaps, after all, when our learned professor comes to the solemn hour of death, he will find it an unspeakable solace if he can only dispense with his hard theory, constructed with such ceaseless thought and exquisite skill, and calmly rest as a little child in the everlasting arms of the Almighty Being, who in the days of his vigour he would only recognize as a cold existence in the universe, unmoved and unaffected by the calamities of His creatures.

Evening is drawing on in the village of Fern Glen; over the hills she glides so gently, closing the eyes of every floweret and feathered warbler, and Eva, as is her wont, takes her way to the church to play the organ, and its deep tones float out upon the summer breeze.

Eva still pursues her profession as indefatigably as ever, and it only requires a glance at some of her later productions to discover how much she has improved. But to see the best specimen of her handiwork you must visit Haverleigh Park, where hanging above the couch, in a certain luxurious study, is a spirited painting of a hunting scene; the glowing colours of the equestrians, and the grey mist, which lies like a curtain over the sodden grass in the background, are admirably managed, and every touch evinces the artist's hand and eye.

Alice is just the same, and her calm sweetness of disposition is perhaps even more noticeable. As she stands by the ancient ivied porch of the old church, and listens to the sweet sonata as it rises and dies on the quiet night, her whole soul goes forth to meet it, and it seems to her in that hour of inspiration that *this* is the land of shadows and *beyond* is the land of light.

Roger has not gone to sea after all; he is in Canada, studying engineering, and promises to excel in his profession. His welcome epistles are perused many times over, and their arrival is a red letter day in the quiet country home.

Meeta is a frequent visitor at Haverleigh Park, and to James, who is often a prisoner on account of his delicacy, her coming makes a pleasant change in his uneventful life; at such times she will read to him by the hour. The stories he loves best are of daring and dangerous exploits, such as Captain Marryat delights to narrate to his admiring boy friends.

Mrs. Haverleigh was always glad for Meeta to be with her little son, for the influence of her cheerful and happy disposition roused him and did him good.

Ronald is on the deep seas of the Atlantic, and fond hearts are getting anxious and eager as they wait for his return.

In the deepening dusk the glow-worm with its blue phosphorescent spark shines here and there upon the banks, the lights in the lowly cottages go out one by one, the winds are sleeping low in the valley, and everywhere reigns the perfect stillness of approaching night.

THE END.

HODDER & STOUGHTON'S
GIFT BOOK CATALOGUE,

With Specimens of Illustrations.

BY THE

FOLLOWING POPULAR WRITERS.

W. M. Thayer.	*Mrs. O'Reilly.*
W. H. G. Kingston.	*Phœbe J. McKeen.*
Dr. Macaulay.	*J. B. de Liefde.*
Mrs. Prentiss.	*A. Macleod, D.D.*
Miss Doudney.	*H. G. Adams.*
Dr. Gordon Stables.	*Jacob Abbott.*
Pansy.	*Gustav Nieritz.*
L. T. Meade.	*Edwin Hodder.*
M. A. Paull.	*Sarson C. Ingham.*
L. G. Sèguin.	*J. R. H. Hawthorn.*
Mrs. Reaney.	*Mrs. Webb.*
Marie Hall.	*Author of "The Bairns."*
A. E. Barr.	*Isaac Pleydell.*
W. H. Davenport Adams.	*Etc., Etc.*

London:

HODDER AND STOUGHTON,
27, PATERNOSTER ROW.

I.

FROM POWDER MONKEY TO ADMIRAL. A Story of Naval Adventure. Eight Illustrations. Handsomely bound. 5s. Gilt edges:

> "This story is equal to any ever written by its author. It is full of life and adventure. In the stirring days of the French War, Bill Rayner was a London Arab, and the story tells how by good conduct and bravery Bill rose from the position of powder monkey to that of Admiral. He is, early in his career, wrecked and taken prisoner, with another lad his own age, and the pair go through some exciting adventures before they are again afloat. No detailed description is possible of a story which is from end to end crowded with adventure and incident."—*Standard.*
>
> "Kingston's tales require no commendation. They are full of go. All lads enjoy them, and many men. This is one of his best stories—a youthful critic assures us his very best."—*Sheffield Independent.*

II.

JAMES BRAITHWAITE, THE SUPERCARGO. The Story of his Adventures Ashore and Afloat. With Eight Illustrations, Portrait, and Short Account of the Author. Crown 8vo, gilt edges, handsomely bound, 5s.

> "Few authors have been more successful than the late Mr. Kingston in writing for boys. Possessing a keen eye for dramatic effect, he plunged at once into the heart of his story, and roused the interest of his readers by rapidly unfolding a succession of moving incidents. In 'James Braithwaite' the supercargo's exploits at sea during the early part of this century appear as fresh and vigorous as though they were described yesterday. It is a healthy, hearty, enjoyable story."—*Daily Chronicle.*

III.

JOVINIAN. A Tale of Early Papal Rome. With Eight Illustrations. Cheap Edition. Fcap. 8vo, 2s. 6d.

> "It is a powerful and thrilling story of the early part of the fourth century, when Christianity was rapidly gaining a nominal ascendency in the proud city of the Cæsars, and pagan Rome was giving place to papal Rome. The description of the capital of the ancient world, of the intrigues and corruptions of decaying Paganism, of the struggles and conflicts of the early Church, of the wonderful catacombs in which they found protection and safety in life and a resting-place in death, is singularly graphic, and indicates a skilful and practised pen."—*Methodist Recorder.*

London: Hodder and Stoughton, 27, Paternoster Row.

Specimen of the Illustrations.

BY W. H. G. KINGSTON—(continued).

IV.

HENDRICKS THE HUNTER; or, The Border Farm. A Tale of Zululand. With Five Illustrations. Crown 8vo. Handsomely bound in cloth. Gilt edges, price 5s.

> "'A Tale of Zululand.' A very appropriate region to take a boy in at this time. We have nothing to do, however, with the late war, though there is plenty of fighting before Hendricks' tale is told. The illustrations are quite in the spirit of the book. No one who looks at the frontispiece but must turn to page 201 to learn the issue of the startling scene depicted."—*Times.*
>
> "A delightful book of travel and adventure in Zululand. The reader is introduced to Cetewayo as a young man, before he had attained his rank and pre-eminence as 'a noble savage.' The book is very interesting, and comes appropriately at the present moment."—*Athenæum.*
>
> "A boy may be happy all day with Mr. Kingston's 'Hendricks the Hunter.'"—*Saturday Review.*

V.

CLARA MAYNARD; or, The True and the False. A Tale for the Times. Ninth Thousand. Crown 8vo, cloth, 3s. 6d.

> "An admirable story, in which the mischievous results of Ritualistic teaching are effectively shown. Mr. Kingston has very skilfully introduced some effective arguments against High Church and Romish principles, which serve to make it extremely useful."—*Rock.*

VI.

PETER TRAWL; or, The Adventures of a Whaler. With Eight Illustrations. Handsomely bound in cloth. Crown 8vo, gilt edges, price 5s.

> "In this volume of adventures amongst the icebergs and walrusses, Mr. Kingston has ventured upon what seems to him an inexhaustible subject of excitement to the boyish mind. Here will be found shipwrecks and desert islands, and hair-breadth escapes of every kind, all delightful and spirit-stirring, and all ending in a happy return home after the tossing to and fro endured by Peter Trawl, who will be treasured up in many a boy's memory, and cherished as dearly as the image fixed of Robinson Crusoe."—*Court Journal.*
>
> "A whaling story by the late Mr. Kingston, and promises well. It is a manly sort of book, with a good deal of information in it, as well as the adventures which boys love. A true story of a gallant skipper, who, with the assistance of a ship's carpenter, amputated his own leg, is amongst the notable occurrences recorded."—*Athenæum.*

London: Hodder and Stoughton, 27, Paternoster Row.

Specimen of the Illustrations.

Specimen of the Illustrations.

Specimen of the Illustrations.

I.

GREY HAWK: Life and Adventures among the Red Indians. An Old Story Retold. Eleven Illustrations. Handsomely bound, gilt edges, 5s.

"The editor of the *Leisure Hour* having come across a romantic story of real life, has worked it up into a genuinely interesting Indian story. The illustrations and handsome style in which the book is got up make it very suitable for presentation."—*Sheffield Independent.*

"We cannot better testify to its absorbing interest than by saying that we have read every word of it. It is a unique picture of Indian life and customs —of a state of things which already has well-nigh passed away. It is as instructive as it is romantic. As a book for boys, and not for them only, it can scarcely be surpassed."—*British Quarterly Review.*

II.

ALL TRUE. Records of Peril and Adventure by Sea and Land— Remarkable Escapes and Deliverances—Missionary Enterprises and Travels—Wonders of Nature and Providence—Incidents of Christian History and Biography. With Twelve Illustrations. Crown 8vo, cloth gilt, 5s.

"'All True' contains records of adventures by sea and land, remarkable escapes and deliverances, missionary enterprises, etc. ; is as entertaining as the majority of such books are depressing, and may be welcomed as a welcome present for children. The illustrations are above the average of those vouchsafed to us in children's books."—*Spectator.*

III.

ACROSS THE FERRY: First Impressions of America and its People. With Nine Illustrations. Crown 8vo, cloth, price 5s.

"Dr. Macaulay not only records his own impressions, but he incorporates with them much of the useful and interesting information which an intelligent traveller not only picks up, but takes special pains to furnish himself with. The volume is a series of photographs of America as it was in 1870, and is full, therefore, of practical interest."—*British Quarterly Review.*

HOW INDIA WAS WON BY ENGLAND UNDER CLIVE AND HASTINGS. By B. W. SAVILE, M.A. With Twelve Illustrations and Map. Crown 8vo, price 5s.

"It may be numbered with the books of which Dr. Johnson laments that there are so few, and we may say of it that we regret it is not longer. Mr. Savile writes in a fluent, easy style, and this book is replete with interesting information on the present as well as the past condition of India."—*Record.*

London: Hodder and Stoughton, 27, Paternoster Row.

Specimen of the Illustrations.

BY L. T. MEADE, Author of "Scamp and I," etc., etc.

I.

HOW IT ALL CAME ROUND. With Six Illustrations. Handsomely bound, price 5s.

"It is a charming story. The characters are excellently drawn."—*Standard.*
"The story is worthy of the highest praise. Two heroines divide the interest between them, the 'poor Charlotte,' daughter of the widow whom her stepsons have so strangely deceived, and the 'rich Charlotte,' who, knowing nothing of the wrong, is in enjoyment of the ill-gotten wealth. Our sympathies waver between these two; and Miss Meade's art shows itself to the best advantage in the suspense which she thus contrives to maintain. The 'poor Charlotte's' unworldly husband, too, is a very fine character. Altogether, this is one of the best stories of the season."—*Pall Mall Gazette.*

II.

HERMIE'S ROSEBUDS, and other Stories. With Illustrations. Handsomely bound, price 3s. 6d.

"A collection of short pieces by this gifted authoress, illustrative of the quickening and ennobling influence exerted even on the worst of men by children. The whole series is a powerful and pathetic illustration of the text, 'A little child shall lead them.' 'The Least of These' is a capital sketch, so is 'Jack Darling's Conqueror.'"—*Freeman.*

BY MISS M. A. PAULL, Author of "Tim's Troubles," etc.

I.

FRIAR HILDEBRAND'S CROSS; or, The Monk of Tavystoke Abbaye. With Frontispiece. Crown 8vo, cloth, 5s.

"The volume is beautifully written, and never were the struggles of a true and faithful heart more touchingly depicted. The tenderness of the sentiment which binds the friar to Cicely is depicted with such exquisite refinement and delicacy that many a bright eye will be dimmed with tears in the perusal."—*Court Journal.*

II.

THE FLOWER OF THE GRASSMARKET. With Five Illustrations. Cheap Edition. Crown 8vo, cloth, 3s. 6d.

"There is a healthy moral tone of a very high order sustained throughout the work, and an easy grace and diction, which make it highly commendable."—*Edinburgh Daily Review.*
"A handsomely got-up volume. The story is admirably written. The reader never loses interest in the fortunes of the various characters in it."—*Sheffield Independent.*

London: Hodder and Stoughton, 27, Paternoster Row.

Specimen of the Illustrations.

Specimen of the Illustrations.

15

Frontispiece.

I.

EPHRAIM AND HELAH. A Story of the Exodus.
Eighth Thousand.   Crown 8vo, cloth elegant, 5s.

"Mr. Hodder gives a vivid description of the daily life of the Hebrews immediately at and before the time of the coming of Moses. The picture is full of interest."—*The Queen.*

II.

TOSSED ON THE WAVES. A Story of Young Life.
Fifteenth Thousand.   Fcap. 8vo, cloth, 3s. 6d.

"We cannot think that a boy could take up the book without feeling its fascination, or without rising a better lad from its perusal. The scenes of life on the sea and in the colonies are peculiarly attractive."—*British Quarterly Review.*

III.

THE JUNIOR CLERK. A Tale of City Life. Fourteenth
Edition.   Crown 8vo, cloth, 2s. 6d.

"Mr. Shipton observes that the author described this tale to him as a fiction. He remarks: 'It may be so to him, but for every one of his statements I could supply a fact. It is not merely true to nature as a narration of the means by which young men may be—it is a true record of the ways in which many have been, and many still are being—led to dishonour and ruin.' Such a recommendation as this will be sufficient to ensure for this little book a hearty welcome from many readers."—*Christian World.*

THE WHITE CROSS AND DOVE OF PEARLS. A
Biography of Light and Shade.   By the Author of "Laura Linwood," etc. Sixth Thousand. Crown 8vo, cloth, 5s.

"'The White Cross and Dove of Pearls' will not disappoint the expectations of those who may already have formed justly high opinions of this strikingly original and sympathetic writer's ability to interest, to amuse, and to elevate her readers. It is a fiction without false sentiment, without unhealthy imagination, and without a single vulgar or frivolous idea."—*Daily Telegraph.*

THE SISTERS OF GLENCOE; or, Letitia's Choice.
By EVA WYNNE. Twentieth Thousand. Crown 8vo, cloth elegant, price 5s.

"Its life pictures are skilfully drawn, and the most wholesome lessons are enforced with fidelity and power."—*Temperance Record.*
"An admirable story, illustrating in a most effective manner the mischief arising from the use of intoxicating liquors."—*Rock.*

London: Hodder and Stoughton, 27, Paternoster Row.

Specimen of the Illustrations.

Specimen of the Illustrations.

I. **JUNO AND GEORGIE.**
II. **MARY OSBORNE.**
III. **JUNO ON A JOURNEY.**
IV. **HUBERT.**

With Frontispiece. Fcap. 8vo, cloth, price 1s. 6d. each.

"Well printed and elegantly bound, will surely meet with a hearty welcome. We remember the delight we took in them years ago, and how lessons which they inculcated have left their traces until this day. Dr. Arnold, of Rugby, was one of the warmest admirers of the author of 'The Young Christian,' and recognized in him a man of congenial spirit. For strong common sense, knowledge of child nature, and deep religious fervour, we have had nothing superior to these four delightful stories."—*Freeman.*

"The author of 'The Young Christian,' though his name may not be included in dignified histories of literature, is really an English classic. One of his little books exerted such an influence on Frederick Robertson of Brighton, that its perusal formed a turning-point in the life of that great preacher; and there have probably been thousands on both sides of the Atlantic similarly affected by the writings of the same author. We therefore welcome with peculiar satisfaction the elegant edition of four of his best stories. The four little volumes stand a second reading, after an interval of many years, better than some works of greater pretensions; and in their spiritual motive there is a perennial power."—*Christian Leader.*

SHORE AND SEA: Stories of Great Vikings and Sea Captains. By W. H. DAVENPORT ADAMS. Ten illustrations. Handsomely bound. 5s. Gilt edges.

"A book which is as thrilling as any romance.'—*Scotsman.*

"An interesting book for adventure-loving boys. It contains a capital description of the life, customs, and manners of the Norsemen, together with much pleasantly told information concerning 'Sebastian Cabot,' 'De Soto,' 'The Early Colonizers of Virginia,' 'Drake,' 'Hudson,' and 'Henry Morgan.' This collection will be deservedly popular."—*Pall Mall Gazette.*

"This is a carefully written and thoroughly good book. Mr. Adams has tried to sketch the lives of famous sea captains with fidelity as well as with graphic power. . . . It is the romance of the sea as it has been actually realized, and boys will find it as instructive as it is interesting."—*British Quarterly Review.*

DAVID LIVINGSTONE: The Story of his Life and Labours; or, The Weaver Boy who became a Missionary. By H. G. ADAMS. With Steel Portrait and Thirty Illustrations. Fifty-seventh Thousand. Crown 8vo, cloth, 3s. 6d.

"An admirable condensation of 'The Story of the Life and Labours of Dr. Livingstone.' Comprehensive in range, abounding in detail, and vividly presenting the graphic description of the great explorer himself."—*Record.*

London: Hodder and Stoughton, 27, Paternoster Row.

Specimen of the Illustrations.

CLUNY MACPHERSON. A Tale of Brotherly Love. By A. E. Barr. With Six Illustrations. Crown 8vo, 5s.

CAPITAL FOR WORKING BOYS. Chapters on Character Building. By J. E. M'Conaughy. Crown 8vo, ·cloth, price 3s. 6d.

CONTENTS.—Choice of Occupation—Small Beginnings—The Life Purpose —The Best Capital—Ingrained Working Habits—Business Education— Tent Mates—On Time—Habits of Economy—A Courteous Manner—Weights —Reefs—Second Thoughts—Success out of Hardship—Manly Independence —A Straight Course—Boys who Read—Decision of Character—Conversation —Letter Writing—The Best Praise—The Rest Day—Power to Work—The Tree by the River—By Path Meadows—Enduring Riches—Home.

THORNTON HALL; or, Old Questions in Young Lives. By Phœbe J. McKeen. Crown 8vo, nicely bound, price 3s. 6d.

"An interesting and well-written story. The characters of the girls are well drawn, and the tone of the book excellent throughout."—*Church Sunday School Magazine.*

THEODORA CAMERON. A Home Story. By the same Author. With Five Full-page Illustrations. Seventh Thousand. Crown 8vo, cloth, price 5s.

"A healthy interest, full of the kindly and sanctifying influences of home, breathes through every page. The work is excellent in tone and style. Every girl and boy must benefit greatly by reading such a good and interesting home story."—*Daily Chronicle.*
"They might have stepped bodily out of one of Miss Yonge's books, so carefully are the characters of all the children drawn. The mother is a beautiful character, and the father almost equally so, while the children of such parents could hardly fail to be interesting."—*Court Circular.*
"A pretty story of the great civil war, which though issued in a single volume, comprises not less matter than an ordinary novel, and introduces the reader to many varieties of character, and numerous stirring scenes in the home and on the battle-field."—*Daily News.*

THE WINTHROP FAMILY. A Story of New England Life Fifty Years Ago. By the Author of "May Chester," etc. Crown 8vo, cloth, 3s. 6d.

"A very dainty, winsome volume."—*Freeman.*
"Primitive New England life, hospitality, and home-heartedness are finely wrought out in it. There is a quiet, easy grace, a pleasant sparkle, and a genial attractiveness in the style which exactly suits the life, manner, and personages of the narrative. A most admirable one for home interest and delight."—*Golden Hours.*

London: Hodder and Stoughton, 27, Paternoster Row.

THEODORA CAMERON.

Specimen of the Illustrations.

25

I.

LAUNCHING AWAY; or, Roger Larksway's Strange Mission. With Frontispiece. Crown 8vo, cloth, gilt edges, 5s.

"An excellently written book of incident and adventure mainly in Australia. The author knows how to make such a book interesting, and he has in this one eminently succeeded."—*Scotsman.*

II.

THE PIONEER OF A FAMILY; or, Adventures of a Young Governess. Second Edition. Crown 8vo, cloth, 5s. With Frontispiece.

"Few stories have such an air of reality about them. Mr. Hawthorn has the faculty of drawing his characters in such graphic fashion, that we seem to have known them, and are forced to sympathise with their joys and sorrows."—*Aberdeen Free Press.*

"This book is full of terse and powerful sketches of colonial life, especially as it was seen a generation ago."—*Freeman.*

JOSE AND BENJAMIN. A Tale of Jerusalem in the Time of the Herods. By Professor F. DELITZSCH, Leipzig. Translated by J. G. SMIETON, M.A. Elegantly bound, crown 8vo, 3s. 6d.

"A beautiful story, both in conception and execution. It has an especial value as the work of a renowned scholar and Orientalist. From his minute knowledge of the matters of which he writes, the local colouring which the professor introduces is full and rich."—*Watchman.*

BELL'S STANDARD ELOCUTIONIST. Principles and Exercises. Followed by a copious Selection of Extracts in Prose and Poetry, Classified and Adapted for Reading and Recitation. By D. C. and A. M. BELL. New and greatly Enlarged Edition. Containing over 500 of the choicest Extracts in the English Language, with the Principles of Elocution fully stated. Strongly half-bound in roan, crown 8vo, 544 pages, price 3s. 6d.

"This is the best book of the kind."—*Bookseller.*

"Has long been accepted and held as one of the best books on the subject."—*N.B. Daily News.*

OLIVER WYNDHAM. A Tale of the Great Plague. By the Author of "Naomi; or, The Last Days of Jerusalem," etc. Fifteenth Thousand. Crown 8vo, cloth, 3s. 6d.

"The chief merit of the book is the exquisite delicacy with which it illustrates Christian feeling and Christian principle in circumstances the most trying and varied."—*Weekly Review.*

London: Hodder and Stoughton, 27, Paternoster Row.

Frontispiece.

Frontispiece.

I.

THE BAIRNS; or, Janet's Love and Service. With Five Illustrations. Thirteenth Thousand. Crown 8vo, cloth elegant, 5s.

"A special interest attaches to 'The Bairns.' The characters are forcibly delineated, and the touches of homeliness which seem almost peculiar to our northern kinsfolk impart a peculiar charm."—*Record.*

II.

FREDERICA AND HER GUARDIANS; or, The Perils of Orphanhood. Cheaper Edition. Crown 8vo, cloth, 3s. 6d.

"An exceedingly well told story, full of incidents of an attractive character. The story will be admired by all thoughtful girls."—*Public Opinion.*

"A sweet, pure, and beautiful story, such as may be put with confidence into the hands of any English girl."—*Sheffield Independent.*

III.

THE TWA MISS DAWSONS. Crown 8vo, cloth, price 5s.

"We gladly welcome a new book by the author of 'The Bairns.' That charming Canadian story opened a new field for readers of fiction. The present story is limited to Eastern Scotland. It is a family picture, settling down chiefly to the experiences of a charming old maiden aunt—a most admirable delineation—and an equally charming niece."—*British Quarterly Review.*

LINKS IN REBECCA'S LIFE. An American Story. By PANSY. With Frontispiece. Handsomely bound in cloth, 5s.

"By one of the ablest and sprightliest of American story-tellers."—*Christian.*
"We should like to see every young lady of our acquaintance fully engrossed in the reading of this book. It is an admirable five shillings' worth."—*Sword and Trowel.*

YENSIE WALTON. An American Story. By J. R. GRAHAM CLARK. With Frontispiece. Crown 8vo, cloth, 5s.

"In tone and spirit, plan and execution, this is a superb story. Rich in delineation of character, and in descriptions of real experience. A more fascinating and inspiring picture of a school-mistress, in one prolonged, prayerful, and sustained endeavour to lead an orphan pupil to Christ, was never drawn."—*General Baptist Magazine.*

London: Hodder and Stoughton, 27, Paternoster Row.

Specimen of the Illustrations.

31